Popularizing Science

Communication, Environment, Science and Society Set

coordinated by
Andrea Catellani and Céline Pascual Espuny

Volume 2

Popularizing Science

The Complex Terminological Interactions between Scientific and Press Discourses within the Field of Agroecology

Hélène Ledouble

ISTE

WILEY

First published 2024 in Great Britain and the United States by ISTE Ltd and John Wiley & Sons, Inc.

ISTE Ltd
27-37 St George's Road
London SW19 4EU
UK

www.iste.co.uk

John Wiley & Sons, Inc.
111 River Street
Hoboken, NJ 07030
USA

www.wiley.com

Library of Congress Control Number: 2023952525

British Library Cataloguing-in-Publication Data
A CIP record for this book is available from the British Library
ISBN 978-1-78630-712-5

Contents

Foreword

The book written by Hélène Ledouble, *Popularizing Science: The Complex Terminological Interactions between Scientific and Press Discourses within the Field of Agroecology*, is the result of the author's experience, as a linguist, in an interdisciplinary research project that began in 2017 at the initiative of INRAE (Institut national de recherche pour l'agriculture, l'alimentation et l'environnement – French National Research Institute for Agriculture, Food and the Environment).

It is original in more ways than one: by the questions it addresses (the popularization of science); by the field it concerns (agroecology); by the methods it presents (linguistic corpus analysis and cognitive analysis) and more generally, by the questions it raises about the possibilities of transmitting scientific knowledge to the general public. A crucial (and fascinating) problem underlies the book – is it possible to popularize science? As the author reminds us, epistemologists, educators and, of course, linguists have taken up this question and put it in another way – are scientific concepts accessible to the general public? Isn't popularization a distortion of science? Is it a simple translation? What is the relationship between popularization and education? More generally, how do political and social perspectives interfere with the goal to popularize? Hélène Ledouble introduces these questions in a manner that is highly relevant, relying on an abundant bibliography. Taking the stance that popularization is possible, she bases her reflection on the study of the media coverage of agroecology. This is a field that is not well known to the general public, although it touches on major issues such as food, health, ecology and the climate.

The analysis presented by Hélène Ledouble focuses on the detailed study of several concepts: biocontrol methods, agents and products. The corpus, made up of press articles in English and French, allows us to identify similarities in the way the two languages work. The systematic exploration of the contexts leads Hélène Ledouble to identify processes that may seem contradictory. For example, the significant use of terminological variation (denominative competition), which does not necessarily result in the "simplification" that is assumed to be the primary concern of journalists. This dual tendency (terminological variation versus simplification) underlines a specific construction of meaning, one that is not inherent to the terms but rather developed during the discursive process, involving both journalists and the readers they are trying to inform.

The popularization of any subject through the media brings into play other issues in addition to didactics alone. It is necessary, in one way or another, to spark an interest or even to seduce the reader, who is also a consumer, even if it means putting knowledge on show, sometimes in a way that borders on sensationalism. Hélène Ledouble shows that the use of metaphors (gastronomy, control, conflict), of childish discourse or of specific (binary) arguments to convey a perspective are frequent, and undoubtedly characteristic of popularization as a journalistic genre.

Regarding sensitive issues such as those related to ecology (in one way or another), the reader approaches general press articles with "knowledge" that is more or less accurate, as well as opinions and expectations that can create cognitive biases. Hélène Ledouble describes how, in the specific case of agroecology, a binary framing is implemented, which may introduce incorrect representations. Thus, "organic" and "natural" methods and products are seen as positive, while chemical and traditional methods (and products) are assumed to be negative. In popular imagination, this binary vision has generated negative connotations that are associated with linguistic structures such as the prefix *phyto* (as in *phyto*sanitary products) or the suffix *cide* (as in insecti*cide*), connotations that are not present in the etymology. Journalists must consider the impact of these implicit meanings, in order to try to rectify them without "offending" the readers' beliefs too much. The popularization of a discipline such as agroecology therefore appears to be a complex process that must combine the simplification (of concepts) and the increasing number of terms, collective thought and scientific knowledge, as well as interest from the reader and the transmission of proven facts. Far from being a predefined path from expert to novice, the popularization of scientific knowledge thus appears to be a construction that is developed by the generalist journalist, who hopes to make the reader part of this process, with a necessary effort to interpret science.

However, the journalist's position does raise certain ethical questions. Indeed, on sensitive subjects such as the one dealt with in this book and given the journalists' complex intentions (to inform but also to spark interest), the way in which knowledge is presented can give the reader a distorted vision of the situation, even to the point of feeling that science has become “crazy”. In order to not move too far into sensationalism and emotion, the journalist must maintain a close relationship with the scientific world. Their texts are also situated in a social, political, economic and sometimes local context, which has an impact on how they write. As Hélène Ledouble maintains, the journalist must also clarify the role of the scientist, which is to not only tell a “truth” but to also doubt and share uncertainty, even if that may be unsettling for the reader. In the process of circulating terms, the journalist thus maintains a sort of partnership with the scientists and the readers (and other actors). Together, they participate in the process of popularization and even in the development of the discipline.

The complexity of the popularization process could lead Hélène Ledouble to question the possibility of popularizing via the media. However, her conviction that such an undertaking is possible and necessary leads her to propose alternatives (or rather complementary approaches). She thus evokes the very current work that is being done in various fields to design simplified languages (plain languages). She also suggests assistance that would help the journalist’s writing and the reader’s interpretation by constructing a resource to reflect the different points of view, so as to shed light on all the facets of the problem. We may think that this could mean building a fairly standard terminological resource. However, several aspects would make this resource more adapted to this specific purpose, including links to usage, links to needs or joint construction of the research question. The author thus develops a real plea for the joint involvement of scientists, citizens and mediators in the development of agroecology.

Hélène Ledouble’s work is based on a terminological approach based on corpus. However, by evoking the complexity of the popularization process, it opens up numerous perspectives on the nature of expert knowledge. It therefore raises questions to do with the role of language in developing knowledge, the possibility of a “neutral” approach, the possibilities of disseminating knowledge to non-experts and the role of simplification (linguistic and conceptual) in this undertaking. Of course, a linguistic study in which the reflection combines detailed analysis with proposals for the general public and decision-makers is most welcome. There is no doubt that the author’s reflections, which are anchored in a solid corpus analysis, will not only give food for thought to linguists but also to epistemologists, scientists, students and, of course, journalists.

Anne CONDAMINES
Research Director, CNRS, CLLE (Cognition, Langues, Langage, Ergonomie) and Université Toulouse Jean Jaurès

to the point of feeling that science has become [illegible]. In order [illegible] and sensationalist [illegible] and emotions [illegible]

[illegible]

[illegible] ethical questions, the [illegible] to not only tell a "truth" but to [illegible] interesting for the reader. In the process of [illegible] maintains a sort of partnership with the [illegible] Together they participate in the process of [illegible] development of the discipline.

The complexity of the popularization process could lead [illegible] the possibility of popularizing via the media. However, [illegible] [illegible] is possible and [illegible] [illegible] the problem. We [illegible] [illegible] this resource [illegible] develops a real plea for the popularization [illegible] the development of knowledge.

[illegible]

[illegible] to do with the [illegible]

Preface

By way of a preface to this entire work, and following the example of Jeanneret (1994), we wish to state that, for us, the work of journalists popularizing science is unconditionally "worthy of respect and requires great vigilance", regardless of the form it may take. In this book, therefore, everything is thought out, analyzed and written with a deep consideration for this desire for dialogue, between two perspectives that are divergent by nature – the scientific approach and the media approach – the meeting of which is not only necessary, but a decisive factor for good cross-comprehension. The objective of this book is therefore to try to contribute to the reinforcement and enrichment of this dialogue between science and society.

February 2024

Acknowledgments

First of all, I would like to thank Anne Condamines for proofreading this manuscript before its publication and for writing the foreword. We share the belief that both the linguist and linguistics (applied but also "engaged") have a role to play, in the professional world and in society in general.

I would also like to pay tribute to the late Professor Henri Zinglé, Director of the Laboratory of Linguistic Engineering and Applied Linguistics (LILLA) at the University of Nice, who shared this same vision. He knew how to spark vocations in young researchers thanks to his passion for linguistics and language processing tools.

I would also like to express my gratitude to my colleagues in linguistics at the University of Toulon, for our enriching exchanges and seminars. I would like to thank Michèle Monte in particular, for infusing her energy into our research team, and for her open-mindedness, which ensures the interdisciplinarity that is so essential to scientific research.

Three proofreaders of my manuscript have helped to nourish both the content and the form of my thoughts. Mojca Pecman, professor at Université Paris Cité, has accompanied my work since its outset, and I thank her in particular for her thorough proofreading and precious advice. Aura Parmentier, associate professor at the Université Côte d'Azur, enabled me to enrich this manuscript with her approach to social sciences and her decisive role as coordinator of the Valbio Collective since its inception. I extend my gratitude to Katharine Throssell for her outstanding proofreading and translating skills, unwavering professionalism, and meticulous attention to detail, which significantly contributed to the refinement of (part of) the English version of this book.

I would also like to thank my colleagues at INRAE (notably Thibaut Malausa, Nicolas Ris, and Louise van Oudenhove), and the Valbio and ENI-BC collectives, who devoted time and attention to my work, as well as my colleagues from Université Côte d'Azur and the research team *Transitions*, at the heart of this project.

I am also grateful to Andrea Catellani (Université Catholique de Louvain) and Céline Pascual Espuny (Université Aix-Marseille) for their proofreading, as well as the inclusion of this book in the "Communication, environment, science and society" set, which they initiated.

On a more personal note, I would like to thank all those who read the various texts that led to this book, and who helped improve them through constructive feedback: Vincent, Sophie, Peggy C., Alexandre, David, Dany, Cyril, Isabelle, Caroline, Peggy F. and Karine.

Naturally, my heartfelt thanks go to Astier, who has understood the time I needed for this project to go forwards. In keeping with his initial promise, Astier has always accepted my work as part of our family.

Lastly, I cannot conclude this section without thinking of my parents, to whom I give my warmest thanks for their constant support in my personal and professional life, for our rich exchanges which bring so much to me and for the attention they have always given to my most important writings.

Introduction

For a long time, scientific information has reached the general public through the media. This book will question the public's ability to appropriate or to (re)construct scientific knowledge by reading the national and regional daily press. Originally, the media does not set out to transmit scientific knowledge as its primary objective, and we will discuss the role (and the responsibility) of the media in popularizing science in Part One of this book. Nevertheless, we hypothesize that it is difficult to provide information to the reader without also providing a minimum of knowledge[1]. This poses a major challenge for journalists; they must create high-quality information that is both accessible and intelligible, and which enable the reader to participate – in full knowledge of the facts – in the democratic debate on scientific issues.

In this context, it is important to avoid a simplistic approach to the process of popularizing science or minimizing the complexity of these social practices and technical and semiotic devices[2]. The paradox inherent in this process is now well established, being both "trivial", because it has existed for a long time, and "complex", mostly because it relies on a set of disciplines, and therefore on a set of perspectives. Because the popularization of science relies on the intersection of communication devices, knowledge and social logic, it is of interest to "hard sciences" (chemistry, physics, biology, agronomy, etc.), as well as to the humanities (information and communication, linguistics, history, etc.). Moreover, since the subjects covered by the media concern both the political world and society, the process is of interest to social sciences as a whole (sociology, economics, law, management, etc.).

---

1 We thus follow the principles proposed by other researchers, such as Jeanneret (1994), Véron (1997) or Schiele (2005), for example, who consider that popularizing science requires the dissemination of a minimum amount of scientific knowledge.

2 These terms (*social practices, technical and semiotic devices*) are borrowed from Le Marec and Babou (2004).

In his seminal work *Écrire la science*, Jeanneret (1994) suggested that the popularization project should be taken very seriously, reminding us of this necessary interdisciplinarity, a principle that is now accepted in all research on this subject[3]. The extent and development of the field in recent decades can be seen in the numerous workshops, conferences and publications in this area. They also illustrate the wide variety of approaches that are devoted to it, which we will explore in this work[4].

Our study follows on from research devoted to the "science-society dialogue" and to the dissemination of scientific information beyond experts. The scientific field at the heart of this research is agroecology, and in particular the protection of plants by natural mechanisms (insects or substances of natural origin, for example). Our objective is to study terminological and cognitive issues when popularizing the processes of *biological control* (in French and English)[5]. The analysis of this field is particularly interesting for at least three complementary reasons, which we will outline below.

First, although its denominations and concepts are still not well known to the general public, scientific research on biological control is constantly progressing, and the development of its methods is likely to have a significant societal and environmental impact in the decades to come. Faced with the pressing issue of

3 In the wake of other seminal works such as Schoenfeld (1980) and Nelkin (1986).

4 Among many references, let us mention for example: the conference "Parcours linguistiques de discours spécialisés" (Sorbonne and Cediscor 1992); the workshop "La communication de l'information scientifique" (Centre de recherche en information et communication, 2001); the conference "Le traitement scientifique de la complexité" (Lille, 2001); the conference "Sciences, Médias et Sociétés" (ENS Lyon, 2004); the conference "La publicisation de la science" (Institut de la communication et des médias, Grenoble, 2004); the conference "Sciences et société en mutation" (CNRS, 2007); the workshop "La mise en récit de discours spécialisés" (GERAS, 2015); the workshop "Rencontres, Sciences et Société" (University of Lyon, 2017); the conference "Les nouvelles formes de la vulgarisation et de l'écriture du savoir" (University of Sorbonne Nouvelle, 2018); the ACFAS 2019 congress "Engager le dialogue Savoirs-Sociétés" (University of Montreal, 2019); the workshop "Pesticides, approches pluridisciplinaires" (University of Burgundy, 2021), etc. There are also numerous journals or special issues of journals devoted to these topics: *Public Understanding of Science*; *Les Carnets du Cediscor*; the *Hermès* journal; *Discourse Studies, Discourse & Communication*; the proceedings of the TIA (*Terminologie et Intelligence artificielle*) conferences; *Les enjeux de l'information et de la communication* (2021); "Recherche scientifique et Médias" in the *Revue française de la SFSIC* (2020); *Études de communication*; *Les Cahiers du journalisme*; *Quaderni*, and other books and multidisciplinary research works that we will mention throughout this book.

5 These and other terms related to this specialized area will be defined in Chapter 4, as well as in the glossary at the end of the book.

environmental protection, society is taking up these subjects and – as already noted by Guilbert (1975) – the situation is therefore becoming favorable for the observation of an object of discourse and its evolution.

Second, this field is gaining momentum in the French and international news, as recent regulations (see the law for Agriculture, Food and Forestry[6], the Labbé law[7], the Ecophyto II governmental plan[8], etc.) impose a transition towards more natural methods of plant protection.

Third, the (plant protection) products developed in this field are presented as alternatives to (chemically produced) phytosanitary products (commonly called "pesticides"), which have been disparaged for a long time and are increasingly controversial today. Since the work of Carson (1962), many multidisciplinary studies have been devoted to these controversial products[9], as well as to the necessary transition from agrochemistry to agroecology[10].

Citizens are therefore likely to have a positive preconception of this field, given that scientific research is seen as being able to propose concrete solutions for the reduction of pollution (in soils, waterways, crops), with positive health benefits[11].

On the other hand, agroecology is a complex scientific field that also has controversial aspects, if we look into the processes involved: biological control methods involve human intervention in the natural ecosystem and affect living organisms (macro or micro-organisms). It is therefore likely to evoke other sensitive and widely covered issues that are related to human interference in natural phenomena (genetically modified organisms, cloning, etc.). The way in which these

6 Law No. 2014-1170 of October 13, 2014. See: https://www.legifrance.gouv.fr/jorf/id/JORFTEXT000029573022 [Accessed November 1, 2023].

7 See: https://www.legifrance.gouv.fr/eli/loi/2014/2/6/2014-110/jo/texte [Accessed November 1, 2023].

8 See: https://agriculture.gouv.fr/le-plan-ecophyto-quest-ce-que-cest [Accessed November 1, 2023].

9 Examples include some recent multidisciplinary studies on pesticides: (Roussary et al. 2013; Aulagnier and Goulet 2017; Chateauraynaud and Debaz 2017; Botero 2021; Foucart 2021; Jouzel 2019; Lambert 2020).

10 Among a profusion of multidisciplinary works and publications on these issues, the following references are strongly similar to our work, by their theme or their methodology (see Galochet et al. 2008; Deguine et al. 2016; Chlous et al. 2017b; Catellani et al. 2019; FAO 2019; HLPE 2019; Fauvergue et al. 2020; Allard-Huver 2021).

11 According to the *One Health* concept, or *Global Health*, which addresses the health of humans, animals and plants in a holistic manner. See: https://www.inrae.fr/alimentation-sante-globale/one-health-seule-sante [Accessed November 1, 2023].

subjects are presented by the media may therefore have an impact on how the citizen understands and perceives the field (in a positive or negative light). Indeed, the capacity of any individual to understand, to accept or to reject scientific progress will largely depend on the way in which it is presented (named, designated, described) in the media, and thus popularized.

This is the central focus of this book. Here, we set out to explore the way in which the field of biological control is presented by the written press, through the general perspective and methods of applied linguistics.

The book is divided into three parts, which will enable us to explore these questions in greater depth. Part 1 is made up of three chapters and sets the overall framework in which we place ourselves, that of the challenge assigned to the media in this dialogue between science and society. We address the multiple forms (and perceptions) of this dialogue, the role and the responsibility of the media in this process, as well as the complex relations between scientific and media methods and perspectives (Chapter 1). These elements are then enriched by a text-based approach, dealing with discourses around science and, in particular, the link between language and the different forms of knowledge that can be conveyed (Chapter 2). This first part concludes with the way we envisage textual data analysis, and the presentation of theoretical and methodological tools that are specific to textual terminology and which constitute an interesting entry point for the study of media discourse (Chapter 3).

Part 2 of this book is devoted to the case study at the heart of this research: the media coverage of the specialized field of biocontrol. After a presentation of scientific methods and products involved, and of the dual French and English corpora selected for this study (Chapter 4), the two following chapters highlight the results of the analysis in context. First, we present the different terms that refer to the same methods and products in this field of specialization, in other words, the competitive denominations throughout press articles (Chapter 5). We then discuss the description of these objects of discourse, through various simplification strategies, for example, the use of a "binary" cognitive procedure pitting the biological against the chemical, as well as the use of reasoning by analogy (Chapter 6). These discursive strategies ensure a clear and accessible dissemination of information to a wide audience but can raise questions and even interpretation difficulties for a non-expert reader.

Part 3 of this book consists of two chapters that look back on issues and challenges related to media coverage in the field of agroecology. The first chapter (Chapter 7) presents a summary of the questions generated by the analysis of media discourse, revealing the linguistic instability that can hinder effective communication. The second and final chapter of this part (Chapter 8) identifies a set of challenges that need to be addressed in order to improve communication in this emerging field, which are related to scientific issues in environmental communication more generally.

NOTE.– This study is being conducted as part of an interdisciplinary research project (BOOST: Bioprotection and Biostimulation of Plants) initiated in 2017 by the French National Research Institute for Agriculture, Food and the Environment (Institut national de recherche pour l'agriculture, l'alimentation et l'environnement – INRAE). It is supported by the Université Côte d'Azur (UCA) as part of its structuring research projects and brings together several scientific partners (including the Institut Sophia Agrobiotech (ISA), the Group for Research in Law, Economics and Management (Groupe de recherche en droit, économie et gestion – GREDEG), the URE Transitions (UCA) and the BABEL research laboratory (University of Toulon)[12].

12 The BOOST project (under the IDEX UCA-JEDI of the Université Côte d'Azur), launched in 2017, brings together researchers in biology, agronomy, entomology, chemistry, management, sociology, linguistics, information sciences, psychology, philosophy, etc.

PART 1

The Challenge of Popularizing Science

Introduction to Part 1

This introductory section has various objectives. It begins by presenting the definitions and perceptions of the various notions discussed in this book (popularization of science, science in the media, etc.), before emphasizing their implications for society. In media publications about complex scientific issues, the terminological question is crucial. It represents the link between linguistic forms and specialized knowledge, within discourses targeted to a less expert readership. In this first part, we will explore press discourses on science (Chapter 1), analyze their terminological and cognitive characteristics (Chapter 2) and conclude Part 1 with a presentation of our methods of textual data analysis (Chapter 3).

1

The Dialogue Between Science and Society

The word is there because the problem is there.
The word is awkward because the problem is awkward[1].

Yves Jeanneret (1994, p. 20)

Communicating science to the layperson is a complex subject. There have been many kinds of popular science initiatives since the origin of the "popularizing project" (Schiele 1983). It is therefore important to briefly recall the definition, the denominations and the designations devoted to this project.

1.1. The popularizing project

Popularization can be defined as "making something (that is difficult to understand) easily understood by ordinary people"[2]. When applied to science, it is a complex process that gives rise to a wide variety of mechanisms and materials. Jeanneret summarized them as follows: "Popularization of science is part of a continuum of forms of communication that goes from preprints, reserved for a small number of peers, to the mainstream press" (Jeanneret 1994, p. 176). Between these two opposite poles, there are a range of other materials, such as museum catalogues

---

1 The "awkward word" is the French term "*vulgarisation*" (see explanation in section 1.1.1). Original French version: *Le mot est là parce que le problème est là. Le mot est gênant parce que le problème est gênant.*

2 WordReference Random House Learner's Dictionary of American English © 2023 (https://www.wordreference.com/definition/popularization).

and documentation, training manuals for students, popular science magazines (*Science et Avenir*, *La Recherche*, *The Scientist*, *Scientific American*, etc.) and a number of other media making scientific information accessible to non-experts. Authier described this activity as "conveying scientific information towards the outside, based on knowledge that has already been produced and is circulating within a more restricted community" (Authier 1982, p. 34).

Here, we look back on some of the terms used to describe these practices.

1.1.1. *How we talk about popular science*

Different lexical forms are used to refer to this field of study, without necessarily being equivalent to one another: popularization, divulgation, vulgarization, publicization, scientific communication, science for the layperson, science in the media, scientific mediation, etc. Some terms are favored by specific languages or cultures (such as popularization, or popular science in the English-speaking world, *divulgación* in the Spanish-speaking world). In French, *vulgarisation* is the generic term used for these practices. From an etymological point of view, it is forged from the Latin root *vulgaris*, which is derived from *vulgus* ("the crowd"). Based on this etymology, it has progressively taken on a depreciatory connotation (Jacquet-Pfau 2021). Thus, because of its name and its very purpose, "vulgarized" science has always raised questions in the Francophone world. Some authors see it as an unrealistic undertaking, in the face of an excessively large gap between the world of science and another "ordinary" world. In reading various publications on the subject, it becomes clear that certain designations are tinged with negativity; for instance, they may refer to an artificial culture ("culture de synthèse" in French), an imitation of culture (*"ersatz" or "culture alibi"*), a parody of learning, serving to perpetuate a kind of scientific illiteracy and lack of culture. In describing the position of its detractors, Jeanneret reminds us that in communicating scientific expertise to the layperson, knowledge may lose some of its purity and be considered as degraded (Jeanneret 1994). This passage between two worlds, which some present as a process of translation[3], or even betrayal[4], potentially refers to the shift from a "pure" world to an "impure" world.

Critics claim that the popularization process transforms both the form of knowledge and its substance. Roqueplo (1974) takes up some of Bachelard's (1938)

3 Moles, A. and Oulif, J. (1967). Le troisième homme – Vulgarisation scientifique et radio. *Diogène*, 58, 29–40.

4 In reference to the French expression "belles infidèles" ("beautiful betrayal"), used by the poet Gilles Ménage in 1648, about the translation of a text written by Lucien de Samosate, a Greek writer (second century), by Nicolas Perrot d'Ablancourt (1654).

arguments in the well-known publication *La formation de l'esprit scientifique*, describing the "showcasing" of knowledge isolated from scientific practice. He argues that "it is impossible to truly transmit objective knowledge through the mediation of a discourse/spectacle *alone*, implying a unilateral form of communication and excluding any effective practice"[5] (Roqueplo 1974, p. 220).

Thus, a common critique of popular science is that it "substitutes representations for knowledge" (Schiele 1983, p. 164), and incarnates a "myth of scientificity" (Jurdant 1973, pp. 97–107), insofar as what is transmitted is not "real knowledge".

Nevertheless, considering popular science as sharing scientific knowledge with a non-specialist audience is an ambitious objective, and inevitably leads to a defeatist conclusion (Jacobi 2005). We agree with many authors who wish to "dispel, once and for all, the issue of the scientific illiteracy of citizens" (Schiele 2005, p. 50), to which popular science would be a remediation process. The vocabulary used to describe this process shows how distant it is from scientific knowledge in the true sense (narrative approach to science, showcase approach to science, publicization of scientific issues, etc.)[6]. Popular science, and science in the media even more so, consist of ignoring certain aspects of scientific work in order to "make it understandable as a social activity" (Jeanneret 1994, p. 112).

We now present different perspectives that tend to recognize the virtue of the dialogue between science and society and look at encouraging prospects for the popularizing project.

1.1.2. *Promoting popularization*

Historically speaking, popularization has also been perceived as somewhat virtuous, disseminating a form of "scientific culture" to the common people, so as to compensate for their ignorance "imped[ing] the progress of the Enlightenment" (Jeanneret 1994, p. 45). In the French-speaking world, the term "*culture scientifique*" ("scientific culture") has slowly been replacing the French term "vulgarization" since the 1980s (Raichvarg 2016). It is, however, important to clearly define what is meant by the term "culture" in this phrase, and in the so-called "cultural" approach to popular science. Jacobi reminds us of its ambiguous meaning, whether the approach is anthropological, literary or artistic (Jacobi 2005, p. 70).

5 Original French version: "*Il est impossible de transmettre vraiment le savoir objectif par la seule médiation d'un discours/spectacle impliquant une forme unilatérale de communication et excluant toute pratique effective*" (Roqueplo 1974, p. 220).

6 Wolton (2011) even speaks of "cultural emancipation" (p. 39), illustrating the distance with a real "knowledge-sharing" process.

However, the term *culture* is no less ambiguous when combined with "scientific and technical"[7].

Over the course of the 20th century, this desire to make science accessible to society led to the creation of museums devoted to scientific discoveries (the "Palais de la Découverte" opened in 1937 in Paris), or to "Science centers" (such as the Ontario Science Centre[8], the "Cité des sciences et de l'industrie" in Paris[9], etc.). These facilities innovate by proposing "interactive devices, intended to be both recreational and educational in order to capture, focus and hold the visitor's attention" (Schiele 2005, p. 46). Shiele describes this as the "social project" of scientific and technical culture, the objective of which is not only to (re)enhance the image of science but also to involve the public in the scientific world. It is also a political project, with, for example, the French government launching a series of initiatives in this area since the 1990s. In 2004, the government created a national plan aiming to develop and modernize the ways scientific information is disseminated[10]. In 2012, the French National Council for Scientific, Technical and Industrial Culture (CNSTI) was created, with the ambition of creating an advisory body on these subjects[11]. Two French reports have specifically focused on government initiatives in the area of scientific culture: the OPECST report (2014)[12] and the CNSTI report (2017)[13], the latter being conducted over five years and incorporated into the White Paper on higher education and research.

In order to encompass all of these approaches, Wolton adds an alternative to the list of terms mentioned above, that of *valorization* of scientific research (Wolton

7 This ambiguity can be illustrated by the fact that the emblematic place for making science more accessible to everyone in France (the Cité des sciences in Paris) has been placed under the dual supervision of the ministry in charge of cultural issues and the one in charge of research since 1998 (see Information Report No. 384 (2006-2007) [Online]. Available at: https://www.senat.fr/rap/r06-384/r06-384.html [Accessed November 1, 2023]).

8 See: https://www.centredessciencesontario.ca/parlons-de-nous/qui-sommes-nous [Accessed November 1, 2023].

9 See: https://www.cite-sciences.fr/fr/accueil/ [Accessed November 1, 2023].

10 See: https://www.vie-publique.fr/discours/147667-communique-commun-de-mme-claudie-haignere-ministre-deleguee-la-recher [Accessed November 1, 2023].

11 See: https://www.enseignementsup-recherche.gouv.fr/cid134437/le-conseil-national-de-la-culture-scientifique-technique-et-industrielle.html [Accessed November 1, 2023]; https://www.legifrance.gouv.fr/jorf/id/JORFTEXT000025747676 [Accessed November 1, 2023].

12 Olivier, M. and Leleux, J.-P. (2014). Faire connaître et partager les cultures scientifiques, techniques et industrielles, un impératif. Report, OPECST, Paris [Online]. Available at: https://www.senat.fr/notice-rapport/2013/r13-274-notice.html [Accessed November 1, 2023].

13 See: https://www.vie-publique.fr/rapport/36586-strategie-nationale-de-culture-scientifique-technique-et-industrielle [Accessed November 1, 2023].

1997, p. 41). All of these perspectives have come to public attention thanks to the Unesco Kalinga Prize, which has been awarded since 1951 to the best popular science projects at the international level[14].

A consensus seems to be emerging on the positive mission of popular science, "as a device to educate the greatest number of people" (Jacobi 1986, p. 29). This "non-formal education" is therefore seen as being complementary to traditional education, as was already emphasized in 1958, during a conference on the popularization of science. François Le Lionnais[15] considered it to be more flexible than the teaching of science, and therefore "of national interest":

> Science is moving fast; it is moving faster than education [...]; no matter how good the educational reform is, it will soon be overtaken by science. We therefore need complementary devices and materials. [...] Alongside national education, there must be popular science, which gives teaching more flexibility (Le Lionnais (1958), quoted by Schiele (2005, p. 14))

Both criticized and promoted throughout its history, popularizing science remains, by all accounts, of major importance today. Indeed, its success is in the interest of all those involved in this dialogue.

But who are the actors involved? And who benefits most from popular science initiatives? It could be society as a whole, if we follow Lévy-Leblond (1992), who compares the "level of ignorance" among laypeople to that among scientists in fields in which they are not specialists. This author, himself a scientist, considers that "an expert in one field is a non-expert in almost all others and, as such, is very close to the absolute layperson as far as scientific culture in general is concerned" (Levy-Leblond 1992, p. 17).

Following these positive perspectives on the popularization of science, we will here be focusing on the beneficial or "virtuous" processes whereby a certain form of scientific culture is made available to a very heterogeneous readership, while also contributing to the evolution of science.

We must also question the purpose of this project, particularly as concerns our primary object, the media coverage of scientific issues. What exactly is being popularized?

---

14 See: https://fr.unesco.org/prizes/kalinga [Accessed November 1, 2023].

15 He was then president of the Association of Scientific Writers of France.

In order to answer this, we need to clearly define what is expected of the media. According to Miège, "it is important to make a clear distinction between the organizations that contribute to the dissemination of scientific culture, those that participate in the communication of science, and the 'operators' that are best able to stimulate public debates on scientific questions, which are assumed to be the mass media" (Miège 2004, p. 128).

In fact, in contrast to traditional popular science magazines (such as *The Scientist* and *Science et Avenir*), it is widely documented that people do not read newspapers, whether national or regional, to obtain scientific information or to improve their knowledge in the field of science. In this context, as Fayard specified as early as 1988, "from being a true end in itself, popular science becomes a means, a tool for issues that go beyond the mere scientific content of the message conveyed, and which situates that message in a broader social and economic context" (Fayard 1988, p. 11).

We now turn to the role of the media, particularly daily newspapers, in this dialogue between science and society.

1.2. Science in the media and the multiplicity of actors

Here, we focus on science in the media, i.e. how the daily press covers scientific subjects. The English language has no specific term to refer to the daily press coverage of information, while the French language uses the (polysemous) term "*médiatisation*" (Lafon 2019a; Miège 2019[16]). When dealing with scientific subjects, daily newspapers have a different objective from traditional popular science magazines, but they also need to "popularize" science, to make it accessible to a non-specialist audience. Let us return to the status and role of journalists in this process.

1.2.1. *From scholarly knowledge to social knowledge*

The critical discourse on popular science has clearly identified the limits of this endeavor when the media is given the overly ambitious task of transmitting scientific knowledge. The written press aims to do something slightly different, not to "*reproduce* pre-existing statements of science, but rather to *construct* a discourse on the world, informed by the activity of scholars but of a profoundly different

16 Miège (2019a) identifies four meanings for this term: mediation using technical devices, production-dissemination of content, mediated interactions or multiple informational productions.

nature" (Jeanneret 1994, p. 38). The goal here is not to "make the individual more competent" (Beacco and Moirand 1995, p. 40), but to provide information on issues concerning science and society. The media contribution to the societal debate is thus considered a "more social than scientific explanation" (Moirand 2000, p. 47), where the notion of scholarly (or "intellectual") knowledge is replaced by that of "social knowledge" (Beacco 2000).

The media thus acts as an informant; in order to appeal to its readership, it must ensure that it presents subjects of interest to its readers. Health and environmental issues have long been among the most stimulating themes in the media, as various studies on these subjects have shown (e.g. Miller 1991, 2004, in the United States). Cabedoche (2003) also cites a survey showing that the French are in favor of citizen debates and wish to be consulted on genetic research (38%), modes of agri-food production (38%), GMOs (37%) and energy choices (29%). The citizen's keen interest in science has since been confirmed by various surveys, notably in France by CREDOC[17] in a report published in 2013 on "citizens' scientific curiosity".

When it comes to dealing with these popular themes, Caune specifies that "the recourse to mediation is all the more necessary as some subjects are controversial, and some fields need to improve their conditions of acceptability […], in areas that carry socially recognized risks" (Caune 2008, p. 38).

This can be seen in any number of crises or scandals that have made the headlines, whether they be issues related to food (mad cow disease, avian and swine flu, etc.), the environment (the impact of synthetic chemical products such as DDT, pesticides and glyphosate) or industrial accidents (Chernobyl, Fukushima, etc.). Some of the latter have become "proper-name events"[18] (Krieg-Planque 2009), used as a non-neutral fixed expression that has a lasting impact on how these subjects are perceived in society.

It is also important to emphasize that since 2020, the global health crisis has increased citizens' interest in health and environmental issues, and more generally in all scientific fields related to these themes. When a subject generates significant discursive production, it then becomes a front-page news item and a "discursive moment" (Moirand 2004), likely to appeal to many readers concerned by how these

17 The Centre de recherche pour l'étude et l'observation des conditions de vie (CRÉDOC – The Centre for the Study and Observation of Living Conditions) is a research organization serving economic and social actors in society. See the study, p. 47: https://www.credoc.fr/download/pdf/Rapp/R289.pdf [Accessed November 1 2023].

18 For instance, if we use the word "Fukushima" in media discourse, it refers to all the disastrous consequences of such an accident, without any need to explain further.

matters impact their own lives. In the context of anxiety induced by media coverage, the citizen is thus "less eager for scientific knowledge than anxious about the consequences of 'the scandal' on their health" (Moirand 1997, p. 37)[19].

The media thus plays different roles, providing information, explanation and even reassurance, while also giving a voice to the different actors involved in these subjects. The balancing act is all the more complex as the act of popularizing science also involves the possibility of having to be a whistleblower, in the face of potentially controversial events (e.g. Chateauraynaud and Torny 1999; Chateauraynaud and Debaz 2017).

In order to enable readers to have an overall understanding of a subject (the processes, actors, issues involved, etc.) and to participate in the democratic debate in full knowledge of the facts, the media gives a voice to all the different parties involved by interviewing experts. In this multi-faceted discourse, the journalist becomes the "conductor" of multiple voices, and the "discursive manager" (Moirand 2000) between two worlds: that of science, on the one hand, and that of the presumed audience on the other. In this respect, the role of the science journalist is different from that of a non-specialist journalist who conveys not only the voice of experts but also that of other actors, and who explores the consequences of these "scandals" on society.

We will now delve further into the question of the role and responsibility of the media in the dialogue between science and society, looking in particular at how it gives a place to those considered experts.

1.2.2. *Perspectives on science and experts*

In the early 20th century, the primacy of science was considered a given, as evidenced by the slogan adopted at the Chicago exposition in 1933: "science finds, industry applies, man conforms" (Bensaude-Vincent 2018, p. 27). In many fields, the expert's opinion provides rigor and precision by contrast with laypeople's approximations (Flaysakier 1997; Laberge 1997). Experts also serve to reassure the public in the face of "messages of insecurity transmitted by the media" (Zappalà 1997, p. 183). In recent decades, the situation has changed, however, especially if we consider the image associated with science and the notion of expertise.

---

19 The popularization of science does not only concern emergency or disaster situations but also phenomena that require attention or prevention, for example, the risk of earthquakes, global warming and the impact of the use of technology (and screens) on health.

1.2.2.1. *The evolution of the image of science*

Various scientific fields offer a nuanced perspective, both on the "progress of science" and on scientific discourse in general. Genetic engineering or biotechnology, for example, raise questions about the impact of scientific research on human beings (or living organisms), and about the evolution of society more generally. Alongside this, media reports of issues or scandals involving science have a direct impact on the everyday lives of ordinary citizens. This can be seen in the case of the anti-diabetic drug (*Mediator*) prescribed to treat anorexia which turned out to have serious side effects[20], or the earlier example of the Thalidomide (in the 1960s), a drug marketed as a sedative and against nausea, which proved to be harmful notably for pregnant women and was withdrawn from the market a few years later[21].

In recent decades, doubt has been cast over scientific research and the associated idea of "progress", breaking the original trust people once had in science (Sicard 1997; Wolton 1997; Schiele 2005). Public opinion therefore tends to show a certain wariness (or even mistrust) towards scientific research and is "more aware of the dangers of science, more determined to challenge what are seen as being negative side effects" and now refuses to "accept risks in the name of progress" (Cabedoche 2003, p. 228).

Distrust of scientific discourses emerges as part of a flow of media information that might be considered transitory. But there appears to be a certain "family resemblance" between issues to do with health, food or the environment (Moirand 2007), as they become politicized and focus public attention on issues that are salient for ordinary people – transparency, traceability, the precautionary principle, etc. The connections between such events generate "a space for the construction of collective memory in contemporary society" (Moirand 2007, p.2) liable to result in a loss of trust in science.

This distrust is also due to the fact that citizens are faced with non-consensual, or even contradictory perspectives on controversial topics (for instance, nuclear energy, the contaminated blood scandal in France, or more recently the coronavirus health crisis). Even though some of these subjects have become less controversial, media attention on them has contributed to the emergence of a certain skepticism. Citizens can see that scientific issues may be seen through different perspectives (industrial, social, legal, etc.) and, moreover, are likely to evolve over time. There are various historical examples of such evolutions, such as commonly used industrial products

20 See: https://www.lemonde.fr/societe/article/2021/03/29/affaire-du-mediator-ce-qu-il-faut-savoir-avant-le-jugement_6074806_3224.html [Accessed November 1, 2023].

21 See: https://www.pourlascience.fr/sd/histoire-sciences/la-terrible-affaire-de-la-thalidomide-3343.php [Accessed November 1, 2023].

or popular foods which are suddenly called into question. Recent examples include bisphenol A, which was used in the composition of a large number of food containers before being banned in 2012, or even dairy products which were marketed as "friends for life" and promoted for their nutritional benefits before the 1980s before being challenged (particularly for adults). Another example might be the turnaround on health recommendations in the prevention of sudden infant death syndrome (SIDS).

At the same time, some scientific information reported on in the media is genuinely controversial, as shown by the Sokal hoax (Jeanneret 1998; Beaudet 2000) or more recently the Séralini affair, concerning the toxic effect of genetically modified organisms on rats (Allard-Huver 2020; Domenec 2020). Regardless of the final outcome of a given study, the perspective presented, or the responsibilities of the actors involved, a seed of doubt is planted as to the reliability of scientific research (and the media coverage of it). Given the notion of risk and the fact that knowledge is fluctuating, the role of the media is extremely complex, particularly when it comes to calling on experts. In this recent context of distrust of science, which specialists are likely to speak, to explain or to legitimize the issues under debate? Here, we take a deeper look at these experts, and the evolution of their status in the media discourse on science.

1.2.2.2. *Actors, experts and multiple voices*

There are various figures of "the expert" in the media, and anyone can participate in the debate on scientific questions. An expert may be a recognized researcher, a writer, an activist or an "ordinary person", and "the testimony of the latter, in particular when he presents himself as a victim, is overvalued and immediately given credibility" (Cabedoche 2003, p. 222). The press (especially the regional press) often makes use of accounts that give value to local citizens' voices in the social explanation of events. Additionally, presumed scientific or disciplinary expertise is no longer systematically a mark of credibility. Notoriety, whether it be editorial, media or social, "creates hierarchies more than academic titles" (Delforce 2003, p. 170).

This "discursive heterogeneity (or hybridity)" (Moirand 2014) reflects a shift from a linear model of communication (from the expert to the less expert) via a mediator (in this case the media), to more complex models. Indeed, there is a gradual shift in the relationship between science and society, where more actors are involved. Wolton describes a "multiplication of actors and parameters to be taken into account in a context of communication rather than information, of debate rather than popularization" (Wolton 2011, p. 35). According to the author, we have shifted from two to four actors in these communication processes, and even four logics (science, the public, politics and communication), with a view to "ensuring a minimum of cohesion between necessarily heterogeneous worldviews" (Wolton 2011, p. 42).

The media no longer necessarily relies on specialists to identify and explain the facts, but instead discusses the associated information and puts it into perspective with other events, topics or knowledge. Considering popularization as a type of informal education, Jeanneret (2004) speaks of a "culture of permanent debate [...] in which academic, social, but also experiential knowledge is combined to produce information" (Jeanneret 2004, p. 26). This is all the more important today with the development of digital communication tools, which have given rise to new roles, for example that of the "specialized amateur", within scientific blogs or forums, illustrating the emergence of a horizontal and vertical transmission of knowledge (see Moirand et al. 2016 2016; Reboul 2020).

In order to further explore different perspectives on science and scientific communication in general, we will now address the issue of communication models that attempt to describe these interplays between science and society.

1.2.2.3. *New communication models*

One of the criticisms of the popularization of science partly stems from the initial model, which considered the gap between the knowledge of scientists and that of the public as immutable and intangible (known as the "deficit model"). This "traditional" model is based on the asymmetry between the actors in communication, and on the linear paradigm via the middleman (science on one side, the public on the other and, in the middle, the mediator who acts as a go-between). In this perspective, the public is seen as external and monolithic, and devoid of scientific knowledge (e.g. Gross 1994; Bucchi 1996; Bensaude-Vincent 2010; Brossard and Lewenstein 2010; Secko et al. 2013).

However, the last two decades have seen this model challenged. It has been criticized for the way in which public knowledge is assessed, often based on specific, out-of-context questions on specific scientific points (e.g. questioning the citizen about the rotation of the Earth around the Sun (or is it the other way around?), the time of this rotation, etc.) (Bucchi 1996). This way of assessing knowledge is controversial for several reasons. On the one hand, it gives a limited and momentary picture of the general scientific knowledge of a citizen. On the other hand, it systematically prioritizes questions that are related to natural or physical sciences for instance, thus neglecting other scientific fields that are considered as secondary. Certain researchers thus refute the principle that "hard sciences" are the most important, arguing that other perspectives are equally decisive for the interpretation of scientific facts, taking into account other issues such as society, politics, the economy and ethics (see Burns et al. 2003; Delforce 2003; Priest 2010;

Trench and Bucchi 2010; Secko et al. 2013), or defending more systemic and comprehensive approaches (see Morin 2005).

The posture and the status of the citizen in this process have also evolved. In the vein of Fourquier and Véron (1985), or Albertini and Bélisle (1988), Jeanneret (1994) refused to see simplified science as common (or basic) thought, and took the reader's modes of reasoning very seriously: "Common knowledge is not just incomplete or erroneous thought; on the contrary, it is an effective cognitive training, with respect to its own objectives" (Jeanneret 1994, p. 114). This knowledge, although imperfect, is nonetheless constructed and has social and personal relevance, prioritizing "conclusion over reasoning" (Jeanneret 1994, p. 125). Myers confirmed this position:

> We cannot understand why there are tensions about genetically modified organisms, vaccinations, or climate change if we assume that science is distinct from the rest of culture, and that the public is, on scientific matters, a blank slate (Myers 2003, p. 274).

The author illustrates this principle through the example of parents whose child suffers from a medical condition, and who must confront particular difficulties on a daily basis; he considers that to a certain extent they become specialists or semi-specialists.

Thus, other models of communication are emerging, in which the citizen is more involved in the chain of scientific communication[22]. Raising awareness among citizens, with a view to enabling enlightened participation has been a goal since the 1990s, particularly for environmental issues (Boillot Grenon 2015). This is what "citizen science" or "public participation in scientific research (PPSR)" seek to achieve, through the deliberate and active promotion of citizens' appropriation, and even production, of knowledge (Houllier and Merilhou-Goudard 2016; Catellani et al. 2021). Different models currently co-exist (community-based research, collaborative research, living labs, experiential research, etc.) as a result of the opening of scientific projects beyond professional researchers.

This conception remains increasingly important today, especially with regard to sensitive fields of study such as environmental management and agroecology (Chlous et al. 2017). In this context, the citizen's perspective becomes prominent as soon as it is associated with a personal interest or integrated within a project (Laügt 2001). Thus, the public is no longer simply passive; it is involved in the issues at

22 Although most of these models are based on scientific communication in a general way, as well as on scientific journalism, we will see that they are also of interest in dealing with science in the media in general.

stake, based on its own knowledge and socio-cultural context. The non-specialist may even have a potential for action, being able to assist scientists in their own reasoning (e.g. Bucchi 1996; Logan 2001; Delforce 2003; Leach et al. 2009; Brossard and Lewenstein 2010; Avenier and Cajaiba 2012; Secko et al. 2013). Bucchi (1996) suggests that efforts in terms of negotiating ideas and interests with citizens would even go beyond the scientists' initial intentions[23] (Bucchi 1996, p. 384).

Schiele thus reminds us of the circularity (and therefore, the positive feedback loop) between a positive attitude towards science and technology[24], and a certain amount of knowledge, which lead to a better understanding of scientific issues by the citizen (and, ultimately their capacity to engage in democratic debate) (Schiele 2005, p. 32).

There are also alternative perspectives to the traditional popularization of science, which assumes an intermediary. One of these is the "deviation" model, in which the scientist speaks directly to the public (Bucchi 1996; Cabedoche 2003; Davies 2008), thus modifying the relationships and the nature of the information reaching the target audience. We return to these issues in Part 3 of this book, focusing on the role of journalists as well as on the prior role of scientists and the subsequent role of readers, in moving toward greater public involvement.

In the interconnected challenges the mediator faces, a parallel can be made between these two approaches (science and the media), to show how they diverge as well as what brings them together.

1.3. Scientific and media approaches to popularization

We first address what differentiates the two approaches at the heart of the dialogue between science and society. This will be important for the case study presented in Part 2 of this book.

1.3.1. *Divergence between scientific and media approaches*

The following three points of divergence between the two approaches involve observation and experimentation, doubt and uncertainty, and the issue of time.

---

23 Source: "Efforts should instead be devoted to describing it in terms of an open-ended negotiation of ideas and related interests, and one may expect its consequences to go well beyond scientists' original intentions."

24 In this regard, he cites Miller's study (1983).

1.3.1.1. *Observation and experimentation*

The term "scientific approach" is used here to refer to both the notion of science, which generates a set of authoritative knowledge, and the notion of research, which aims to investigate questions that do not necessarily have stable or immediate answers (Klein 2020a). Even though scientific paradigms vary and evolve according to different fields of science[25], the experimentation inherent in the scientific approach is based on the search for a result that is not immediate, but rather reached through a period of observation and, potentially, trial and error. As early as 1988, Devries noted that some "medical advances took years to come to fruition, and might not have occurred at all, if their initial success rate had been the deciding factor in their development" (Devries 1988, p. 889). As an example, he mentions scientific achievements that made heart transplantation possible, which had initially failed many times. He extends this to many other medical successes such as dialysis, coronary artery bypass surgery and pacemakers, pointing out that these experiments were mostly built on failures before they brought tangible results.

The absence of such experimentation in the process of popularizing science was highlighted by Roqueplo (1974) in his book *Le Partage du savoir*, providing a nuanced image of this process. He suggests that, without scientific experimental practice, a non-specialist cannot appropriate knowledge, he can only construct a representation of it.

However, the press has a more pragmatic perspective, focusing on the impact of scientific discoveries on the citizen's (quality of) life, as noted above. Thus, "what is intended to be a long experimental series is assessed on a case-by-case, quality-of-life basis" (Devries 1988, p. 889). What ultimately prevails in the media process is the notion of useful information for the reader. Calsamiglia (2003) explains the value of the scientific object, differentiating the two approaches as follows:

> For the former (the scientists) the object has an immanent value in scientific and specialist contexts. For the latter (the public) the value is external to all the theories and methods: what is important is its application, its utility and the consequences of its use in people's lives.

25 According to the principles laid down in particular by: Bachelard, G. (1934). *Le nouvel esprit scientifique*. Les Presses Universitaires de France, Paris and Kuhn, T.S. (1962). *The Structure of Scientific Revolutions*, 2nd edition. University of Chicago Press, Chicago.

1.3.1.2. *Doubt and uncertainty*

Science does not aim to "freeze" an object and make it immutable; instead its objects are constantly evolving. Scientific knowledge, which is always in the process of being developed, "is based on doubt, on possible ruptures in the epistemological field" (Sicard 2011, p. 67).

Conversely, media discourse does not rely on hypotheses and doubt: "It needs certainties. He who has the honesty to say 'I don't know' is disqualified. What is needed is an expert who is infallible and adaptable to all situations" (Flaysakier 1997, p. 137). A scientist who has doubts and confides them to a journalist will be discredited even though doubt is the very foundation of his own approach. As Cabedoche confirms, "unless he causes a misunderstanding and runs the risk of facing popular dissatisfaction, it would become difficult for the researcher to show this caution, this humility, this sense of nuance that usually characterizes him in scientific forums" (Cabedoche 2003, p. 218).

This was one of the difficulties encountered during the initial period of the Covid19 health crisis in 2020. Science did not have all the answers to the questions raised by the pandemic (origin, spread, protection, etc.), and furthermore, the medical community did not speak with one voice. Unable to admit a form of uncertainty or ignorance in the face of an unprecedented situation, the media relayed polyphonic[26] information[27]. Although this notion of doubt or uncertainty is acceptable in certain scientific fields, such as astronomy (Moirand 2000), it becomes problematic in other situations when scientists are faced with questions that affect citizens' health. Yet the scientific endeavor as a whole is built on doubt.

1.3.1.3. *Science in the long-term*

Although science advances relatively quickly compared to the education timeframe (see section 1.1.2), it is by nature characterized by a certain slowness, given its own characteristics (observation, experimentation, trials, errors, conferences, publications, etc.). Even though this time has been shortened by the evolution of alternative modes or models of communication (such as the deviation model, see section 1.2.2.3), science's long time frame is a necessary aspect of its approach.

---

26 The notion of polyphony has been studied extensively on various subjects. See, for example, the works of Moirand (1997, 2003, 2007), Jeanneret (e.g. Jeanneret 2000), or Mondada (1995), and the numerous publications on communication about the Covid-19 pandemic.

27 Klein (2020b) calls this the phenomenon "ultracrepidarianism", which refers to giving an opinion on a subject, either by an ordinary citizen or by an expert, about a field in which he is not a specialist.

The press has a different attitude, striving to present new information as fast as possible, ideally in real time. "The media's own timeframe, accelerated by live transmission techniques, is infinitely shorter than that of science" (de Cheveigné 1997). There are numerous problems due to this necessity for speed, and the situation can thus "become dramatic, as in the case of therapeutic drug trials" (Ibid, p. 15). The health crisis, here too, revealed the confusion generated by the uncertainty associated with short time frames, whether it be around the question of hydroxychloroquine or the side effects of the various vaccines. Moreover, the institutional and technical constraints of the media leave little room for a verifiable scientific explanation and, consequently, "there is a great risk of seeing science trapped in a reductionist communication process" (Sicard 2011, p. 150).

Attempting to characterize the divergences between these two approaches, Caune sees in them two distinct cultures "which have not been able to establish links between the work of experimentation and formalization that is constitutive of the scientific theory which constructs facts, and the work of interpretation conducted by social sciences and the media, which analyzes how scientific statements propose a representation of the world" (Caune 2008, p. 46). However, these two cultures do share a form of common content, that of scientific information, thus suggesting points of convergence which we will discuss in section 1.3.2.

1.3.2. *Convergences between these approaches*

Below, we present three important notions that bring together scientific and media approaches: the quality and usefulness of the information transmitted, the multidisciplinarity and complexity of information, and lastly, communication practices.

1.3.2.1. *Quality and usefulness of the information provided*

Beyond the issues of popularization for the general public exposed above, it is important to stress the reciprocal importance of science for the media, and of the media for science. As Devries underlined, the two approaches share a fundamental principle, that of guaranteeing fair and quality information: "Neither the press nor the medical staff can be excused for the release of inaccurate information. All suffer when this occurs and all must bear the consequences, [...] accuracy is non-negotiable" (Devries 1988, p. 886). This observation therefore requires cooperation between actors involved in the process, in order to avoid conveying unofficial information from ill-informed sources, and to ensure that the right information is communicated. While this cooperative principle is not without problems or risks, the absence of cooperation could, according to the author, be even more problematic[28] (Devries 1988, p. 886). The author suggests that popularization of science allows citizens to be better informed about medical issues, in the case of the risk of stroke for example, about heart performance, the benefits of a healthy diet or the lack of physical exercise. Devries even recognizes the "dignity" of the two approaches (scientific and media) when they contribute to the dissemination of qualitative information that is useful to the citizen.

From this same perspective, an American study by the *National Science Board* (cited in Schiele 2005, p. 31) attests to the fact that more citizens now know that antibiotics do not work against viruses, due in part to press coverage of bacterial illnesses and resistance to certain treatments. Similarly, in a recent study of media coverage of pesticides, Botero points out that the information made available in the media is likely to become "knowledge in action, transforming cognitive data into a solution" (Botero 2021, p. 116). A qualitative and efficient dissemination of knowledge becomes relevant in different ways, whether through social acceptability, education or action, thereby fighting against information that could be described as fraudulent or deceitful (Thouin and Lapointe 2016). The circulation of knowledge is at stake, and its popularization also offers the scientist the opportunity to address, in an admittedly constrained space, a specific fundamental point that was previously unspecified or ignored (Lipani and Pascal 2020). This mediation could be assumed to be even more useful and immediate when medical and political problems are associated (Zappalà 1997, p. 188), as citizens are more likely to be interested in topics that are useful in their personal lives (Burns et al. 2003; Priest 2010).

Moreover, the benefits of this communication are important not only for the image of research but also for its funding. Indeed, it is now recognized that national visibility through the media is strategically necessary to ensure adequate financial support for research (see, for example, Nelkin 1986; Devries 1988; Jensen et al. 2008; Rouquette 2011).

The virtuous circle of popularization can thus be considered from an overall perspective. The dissemination of research results leads to a better understanding of the field, which in return benefits from greater financial support given its greater visibility, and which can then develop further. Several studies also underline the positive impact of this process on researchers' careers (e.g. Jensen et al. 2008). Moreover, there are also ethical requirements to inform citizens about the impacts of research, the latter being generally financed by the public sector, and therefore by citizens themselves (Lamontagne 2016; Thouin and Lapointe 2016).

---

28 Source: "Cooperating with the press is more advantageous to everyone concerned than taking the opposite tack. Although cooperation may entail some bother and risk, non-cooperation may well be more troublesome."

Rouquette summarizes the situation in scientific communication as follows: "Without results with tangible implications, without a directly measurable effect on people's daily lives, a scientific discovery has little chance of being publicized" (Rouquette 2011, p. 30). He thus sheds light on the difficulty for fundamental research to be publicized (Rouquette 2011, p. 30), even though it is indispensable as a driving force and guarantor of the quality of applied research.

1.3.2.2. *Multidisciplinarity and complexity of information*

Scientific and technical information is complex in itself. Communicating it outside the scientific field is no less complex. If we consider public health, for example, we are dealing with a multidisciplinary field by its very nature (in addition to the medical and scientific aspect, it encompasses social, economic, sociological and political concerns). This complexity is not conducive to rapid and complete processing of information (Flaysakier 1997). Thus, according to Laügt (2001), "the role of the mediator, and a 'noble' role in that respect, would be to no longer try to reduce this complexity but instead to try to contribute to societal debate around scientific questions, by betting on a communicational strategic action".

Furthermore, researchers themselves potentially have "a plurality of social identities" (Delforce 2003, p. 171), as a citizen (with values and ideals) and possibly also as a member of the medical profession. We are thus "in the presence of multiple truths, held by different social actors, all of which claim to provide adequate and valid explanations of the phenomena involved, as well as legitimate goals for action" (Delforce 2003, p. 169). Jollivet confirms this situation, specifying that "in the researcher lies the citizen" (Jollivet 2007, p. 419). And behind the citizen, there is potentially a user, a victim, or an activist, possessing a form of knowledge of his own.

Lastly, we address a final point of convergence between the two approaches, which concerns communication practices.

1.3.2.3. *Communication practices and the production of knowledge*

Some consider that scientific and media approaches are opposed when it comes to "producing" something – with one producing knowledge and the other not producing it. Nevertheless, other researchers, such as Véron (1997), see them both as productive. He argues that the media "produce the social reality of which they speak" (p. 26). Charaudeau also corroborates this claim: "The media do not convey what is happening in social reality, they impose their construction of public issues" (Charaudeau 2011, p. 12). We are dealing with the production of a social reality that is co-constructed between science and readers, through the intermediary of journalists who obtain information from specialized sources.

This takes us back to the question of knowledge production. Although the journalist does not set out to create knowledge, the same is also true of the scientist. We might assume that every scientist wishes to produce immutable and absolute knowledge able to promote “scientific progress”, which can be rapidly shared with the rest of society. However, this production of knowledge seems to be more the consequence of scientific work, resulting from its fundamental objective which is to observe, understand and document the phenomena under study. According to this principle, it is difficult to criticize scientists for producing knowledge that is not always definitive, and the media for having to transmit a degree of uncertainty in this knowledge.

We must therefore keep in mind that the media approach “above all consists of ensuring encounters between actors, mindsets and legitimacy. It is this intersection of issues that scientific communication is all about.” (Jeanneret 2000). As Caune rightly points out, “any culture, be it scholarly or popular, scientific or literary, cannot be dissociated from the way in which it is circulated, and exchanged” (Caune 2008, p. 46). This claim is a relevant transition towards the question of discourses on science, which is addressed in Chapter 2.

2

Discourses on Science

Is it necessary for me to clarify that in science, as opposed to magic and religion, knowledge is communicated?[1]

Jacques Lacan (1999, p. 357),

Chapter 1 provided an overview of the principles and issues of science popularization. We will now explore these questions in greater depth by focusing on textual material, through the study of discourses and language conveying information of a scientific nature. However, even from the strict perspective of linguistics, discourses can be approached according to different methods (study of the lexicon, argumentative or enunciative strategies, dialogism, etc.). As stated in the introduction, we wish to analyze the relation between language structures and the associated knowledge, in order to report on cognitive issues when reading press articles in a specialist field. This chapter therefore focuses on the simplification of complex scientific questions, using a linguistic and cognitive approach. Let us first return to the complex question of "genres" in the field of scientific communication.

2.1. Genres in scientific communication

There is a wide range of discursive productions illustrating "science being explained" (Mondada 1995, p. 57). They may be categorized from a "topological" perspective, as advocated by Authier (1982), considering both an inside and an outside of the scientific community (see section 1.1). Loffler-Laurian (1983) favors a different categorization, according to the different types of discourses on science

1 Original French version: "*Ai-je besoin de dire que dans la science, à l'opposé de la magie et de la religion, le savoir se communique?*".

(specialized scientific discourse, semi-popularization, popular science discourse, educational scientific discourse, dissertation or thesis, official scientific discourse). Other researchers have highlighted a more linear perspective, such as Jacobi:

> There is not, on the one hand, a scientific discourse which is unintelligible to the average audience, and on the other hand, a reformulation of the first one, intended for the greatest number of people, but a *continuum*, in which the writers, their texts and their diverse intentions are intimately intertwined (Jacobi 1986, p. 22).

This perspective is shared by Cloître and Shinn (1985), who bring together the wide range of forms and functions of popularization under the term *expository science*, defined as "a sort of continuum of methods and practices utilized both within research and far beyond, for purposes of conveying science-based information, whether as pure cognition, pedagogy, or in terms of social and economic problems". This model of continuity is represented in the form of a funnel (see Figure 2.1), expressing both a continuous flow in the communication of science-based information, as well as stages, depending on the levels of specialization. This perspective symbolizes the process of filtering information according to the audience. Expert knowledge becomes less specialized as we move to more "popular" audiences.

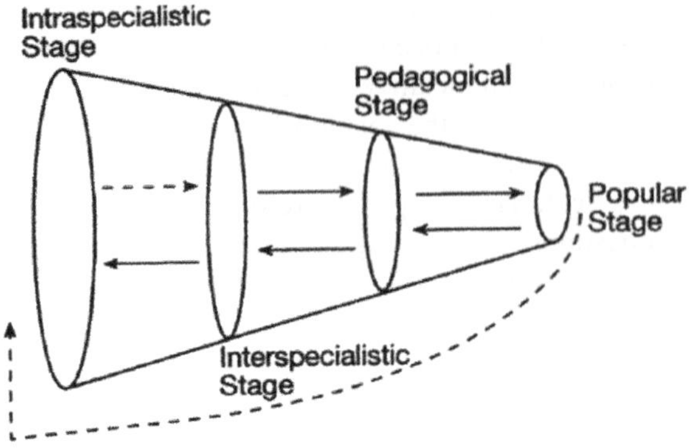

Figure 2.1. *Continuum of methods and practices in science popularization (from Cloître and Shinn (1985))*

Kristiansen and Andersen (2012) offer an alternative representation based on the "product" of this communication, i.e. the differences between textual genres, from scientific articles to daily newspapers (see Figure 2.2).

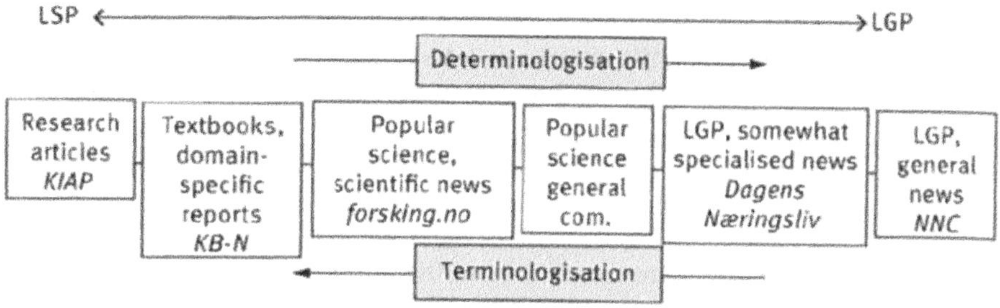

Figure 2.2. *Continuum of discursive productions about science (from Kristiansen and Andersen (2012))*

This diagram, from the most specialized on the left (LSP for Language for Specific Purposes) to the least specialized on the right (LGP for Language for General Purposes), allows us to look more closely at the notion of genre, which is crucial in the study of language and discourse. According to Rastier (2004), any text (oral or written) depends on a genre: one that not only ensures the link between a text and a discourse but also between a discourse and the context in which it was produced. This principle is crucial to the study of terminology, especially when it is based on a textual approach that is necessarily linked to a specialized field.

Addressing the complex question of genres, Pecman and Kübler (2022) recall Swales' perspective that genres are developed by a discourse community in accordance with their collective shared goals, particularly its communication goal. Thus, genres are immediately dependent on the domain to which they refer and:

> it should be recalled that communicative purpose is, according to Swales (1990, p. 10), the prototypical criterion for genre identity. […] Indeed, if we adopt the communicative point of view, genres act as sophisticated models and their role is to facilitate communication by encrypting the purposes and patterns of the communication (Pecman and Kübler 2022, p. 272).

Bathia (2002) illustrates this principle by giving the example of different genres that fall within the same domain, as in the legal framework ("cases, judgments, ordinances, contracts, agreements, etc.").

The complexity linked to genres lies in the fact that they are neither fixed nor immutable, and can also be associated with the notion of "register" (Halliday 1985)[2].

2 According to Halliday (1985, p. 29 and p. 38), a register is "a variety of language, corresponding to a variety of situations".

Registers can then be organized according to different typologies, such as the one defined by Biber (1999) which contrasts conversation, fiction, newspaper language and academic prose. Moreover, "there is generally another meaning attached to the notion of register, in both English and French, which is 'language level' and which refers to a different type of category and set of values, such as literary, formal, neutral, informal, familiar, slang […]" (Pecman and Kübler 2022, p. 272).

The genres and registers are then linked to the notions of terminologization and determinologization, represented in Figure 2.2 by two arrows pointing in opposite directions. In this work, we are particularly interested in this second linguistic phenomenon, which depends on the domain, the genre and the register studied, and describes the shift from LSP to LGP. According to Condamines and Picton, this phenomenon differs from popularization in the sense that "it is not a deliberate process of bringing scientific knowledge to the general public". It is, however, seen as an unconscious phenomenon to both speakers and interlocutors, due to the circulation of terms in specialized fields with a growing public interest (Condamines and Picton 2014, p. 168).

However, this circulation process is not necessarily consensual. While some authors describe a linear and one-way approach, other researchers consider it to be more of a circular procedure (see Ciapuscio 2003; Gülich 2003; Moirand 2003; Humbert-Droz 2021; Picton et al. 2021).

In Part 2, we will return to this question of the term circulation, applying it to the field and genre we are interested in, i.e. the protection of plants by natural mechanisms as seen by the daily press (the register being linked to the media coverage of this specialized field). However, let us first address the terminological question at a more expert level of communication, corresponding to discourse between peers.

2.2. Scientific discourse

If we look into discourses within the scientific community, we can also attest to the existence of different genres that depend on the objectives of communication: a scientific article in a journal or for a conference, an abstract, a book chapter, a book, a thesis, a dissertation, a call for papers, an experimental protocol, etc. (see Pecman and Kübler 2022, p. 269). "The structure of the scientific article, which is highly standardized and organized in sections such as 'materials', 'methods', 'results' and 'discussions', operates a distribution that is different to what could be a report of the laboratory's activity" (Mondada 1995, p. 60). It therefore reflects the "methodical aspect of the descriptions that scientists give of their work" (Mondada 1995, p. 61), and sometimes puts practice and experimentation into the background, as well as the choices or decisions made during the latter, whereas the source of scientific knowledge is prior and complex[3]. Thus, "the textual organization is able to organize that which could seem complex and chaotic in the laboratory" (Mondada 1995, p. 61). The scientific discourse is somewhat independent of the objects it deals with, aiming to keep the possible subjectivity of this process at bay. Jeanneret (1994) recalls the "ideal thesis" of objective knowledge, defined by the epistemologist Popper[4], who called for three distinct worlds: first, that of physical objects, the other of states of consciousness and finally that of objective content. In this perspective, the third world (of objective knowledge) is hence independent from the second one, which depends on communities and researchers, allowing scientific discourse to "limit the subjectivity that would be linked to the thought of the person who constituted it" (Jeanneret 1994, p. 111) by reducing all potential subjectivity.

Moreover, from a formal point of view, scientific discourse uses a specific terminology which, in theory, aims at respecting the principle of univocity (one concept/one term), according to the Wüsterian principles of terminology. Despite this ambition of objectivity, scientific discourse remains complex and heterogeneous, and does not always correspond to this situation of univocal communication. Guespin did declare, as early as in 1991, that "science, like technology, is essentially made up of impure interactions" (Guespin 1991, p. 65), illustrating his approach by an interaction between a researcher and a student trainee in the same laboratory, and through the presentation of their project to a ministry (Guespin 1991, p. 66). Discourses and interactions will thus adapt to the target audience and, as a matter of fact, cannot escape terminological variation. The perception of this phenomenon has therefore evolved over the last few decades: variation is not only attested to by the terminology community but, paradoxically, is valued in certain contexts and for certain purposes. We will explore some aspects of this perspective, which splits from the normative view of terminology and supports the terminological variation as a result of the evolution of science and knowledge.

2.2.1. *Terminological variation in specialized discourse*

Since the early 1990s, some researchers have defined new approaches and theories in terminological analysis, advocating a more descriptive approach to terminology: examples include textual terminology (Condamines 1995, 2005; Pearson 1998; Bourigault and Slodzian 1999; Aussenac-Gilles and Condamines

---

3 However, this remark needs to be qualified according to the disciplines, because the "method" section sometimes includes part of the experiments (and can constitute the longest part of the article), particularly in the "hard" sciences. In the humanities and social sciences, this part's significance varies according to the discipline.

4 Popper, K. (1972). *Objective Knowledge*. Clarendon Press, Oxford.

2007), socioterminology (Gambier 1991; Gaudin 1993), cognitive terminology (Temmerman 2000) or the communicative theory of terminology (Cabré 1998). Early empirical studies on corpora also attest to the extent of variation and the need to consider it in terminology (e.g. Bowker and Meyer 1993; Daille et al. 1996; Condamines and Rebeyrolles 1996; Rogers 1997; Temmerman 2000; Kageura 2002; etc.).

The multidimensionality of concepts – or the question of perspective – is thus emphasized by a community of researchers, following the principle according to which an expert consciously chooses one term amongst others to highlight a specific aspect of a concept (especially an emerging one):

> It [Knowledge] can be multidimensional, since experts' understanding of a conceptual system can depend on their point of view. It may be hard to 'capture', since it is constantly changing, and since emergent knowledge can be incomplete and unclear (Meyer et al. 1992, p. 957).

The notion of classification associated with this principle is used when more than one characteristic can be applied to designate a concept (Bowker 1997; Kageura 2002). These phenomena can generate terminological variation based on synonymy, as shown by Rogers (1997). The author illustrates this with a contrastive analysis of German and English terms in the field of genetic engineering (such as *Krankenheit*, *Erkrankung*, disease, illness and disorder) and shows the importance of the linguistic context (and collocations) in determining equivalents. Numerous additional studies demonstrate the significance of synonymy in LSP, for example, in the field of law, with the co-existence of different useful synonymic forms such as environmental refugees/climate refugees (or environmental/climate/displaced migrants, etc.) (Martinez 2021).

In addition to synonymic variation, there is the question of polysemy, in particular in interdisciplinary situations. The polysemy of the term *satellite* was described by Condamines and Rebeyrolles (1996) as referring to a platform, a mobile object, an artificial body, etc. Using these same principles, in her seminal study of biotechnology terminology, Temmerman contrasted units that adhere to univocity (the "one concept – one term" situation and vice versa), as well as others for which this objective would be neither achievable nor desirable: "in the corpus of life science-related texts we studied, we find evidence that polysemy and synonymy are necessary (functional) and inevitable aspects of terminology" (Temmerman 2000, p. 14). The author testifies that "different terms allow for the expression of shifting perspectives and are not arbitrary" (Temmerman 2000, p. 14), and they thus reject the need for univocity when variation has a functional role within discourses.

Since then, a great deal of research work has explored the phenomena of variation in corpora: Kageura (2002) in an extensive study on documentation; Aymerich et al. (2008) in the field of aquaculture; Dury and Lervad (2008) in the field of energy or Drouin et al. (2017) in a publication illustrating terminological variation in multiple fields (medicine, nuclear, fishing, automobile, law, etc.). Candel (2017) also points to the recent evolution of terminological standardization, which tends to consider usage-based standardization, originally in socioterminology[5]: taking into account the dynamics of a linguistic system, spontaneous and collective term dissemination as well as its regulation capacity. Thus, "terminological variation is increasingly valued in standardization records. These are the opening proposals that are, for example, the fields 'Synonym' or 'Abbreviated form'" (Candel 2017, p. 37).

Research work also investigates the reasons and functions of terminological variation. According to Freixa (2006), the causes for terminological variation can be classified as follows: dialectal (geographical, social or chronological differences); functional (related to the level of language and specialization); discursive (focusing on non-redundancy or expressive needs); cross-linguistic (cohabitation of terms in different languages) or cognitive (differences in conceptualization). Other authors have delved further into the rhetorical or cognitive functions of variation, especially the structural aspects within scientific articles (IMRAD-type articles). Pecman shows the importance of the rhetorical function of variation in relation to neology in Earth and life sciences. She explains that

> variation is concentrated in the Introduction, with an echo in the Discussion. In the remaining parts, the choice of a term is stable throughout. The latent tentativeness in a term choice at the beginning and at the end of the paper appears nevertheless as an intentionally controlled device by which the concept is to be assumed as just being discovered (Pecman 2012, p. 51).

This type of variation is linked to the notion of innovation in science. Through the example of a complex term, in its synthetic form (*hydrothermal plume release*) or developed form (*release of hydrothermal event plumes*), the author can observe the strategic choices related to the notification of newly formed terms or problematized concepts, the second term being reserved for the introductory part, the first as detailed throughout the text (Pecman 2012, p. 54).

Pecman also analyzes the cognitive function of variation in a scientific article, when the use of the expanded form of a term makes its meaning (and formation)

5 The French researcher François Gaudin forged the term "*normaison*" to describe this phenomenon, important in socioterminology.

more explicit (being defined in terms of "terminological unpacking", for example, *Ice Age meteoric waters* which becomes *meteoric waters precipitated during the Ice Age*). In this particular case, "instead of just announcing an emerging concept, variation can serve as a support for building knowledge" (Pecman 2012, p. 21). What could be considered as some form of instability in LSP is proven to be relevant for rhetorical and cognitive reasons, not least when it comes to introducing and manipulating new terms and concepts that represent scientific complexity and innovation (Humbley 2018; Pecman 2018).

Before moving on to the characterization of this variation when the discourses circulate outside the scientific field, let us briefly return to the question of the language used in this endogenous communication, as well as to the nature of this language: is it a language specific to each genre within scientific communication? Can we establish specificities, or even similarities, according to the genres and registers? Can we envisage boundaries between LSP and language used to communicate scientific information to a wider audience, other than our peers? These are some of the questions addressed in section 2.2.2, and which will be relevant in our following study.

2.2.2. *Specialized and general language*

LSP is commonly defined in comparison with LGP, in such a way that a dualistic view of the two types of languages has developed over time. An alternative, less segmented approach has been repeatedly demonstrated, establishing the continuity between general and specialized language. Alongside other researchers, we refute the notion of a (marked) boundary between the two, or the independence of LSP and LGP. As stated by Lerat, "the theoretical foundations of a linguistic approach to specialized languages can only be drawn from general linguistics". According to the author, LSP "refers to the linguistic system for the expressive part and to the professional world for knowledge" (Lerat 1995, p. 12). Other approaches consider the existence of sub-languages in LSP, such as the Harrissian perspective. However, by drawing a precise comparison between these two approaches, Condamines shows that they ultimately converge on various points:

> Regardless of how technical they may be, specialized texts always maintain an undeniable kinship with a reference language: the syntactic architecture, most of the function words and quite a few lexical elements make it very clear which language is at work (Condamines 1997, p. 181).

While being produced "within a particular domain of knowledge, and given a socio-professional context" (Condamines 1997, p. 175), LSP relies both on grammatical words to communicate and on other lexical items borrowed from ordinary language. Resche corroborates and substantiates this statement, pointing out that "one must be careful not to be categorical about what differentiates LSP from LGP" (Resche 2001, p. 45), for "the frontier between the two is not as strict as one might think, and languages are not entirely dissimilar" (Resche 2001, p. 41). This situation is even more evident in interdisciplinary situations, as illustrated by the author within the field of environmental economics, in which LSP and LGP are intermingled. Specialized texts and discourses thus have a direct link with general language; they emerge from it and result from its specialization, to serve communication in very diverse fields of expertise. LGP would thus be incorporated into LSP itself within specialized discourses, or what Véron calls "endogenous intrascientific and interscientific communication" (Véron 1997, pp. 29–30): *endogenous* in the sense that it is characterized by an "equivalence" between the status of the writer and that of the reader (both are scientific).

If terminological variation is confirmed at the level of specialized discourse, we will continue this section by looking more closely at the issues of language and variation when the discourse is aimed at a reader without specialized or expert knowledge in the field. In the wake of the works of Guespin (1991) and Gambier (1991), we are aware that the study of the terminology of a domain can only be undertaken by studying the different discourses devoted to science:

> Rather than seeing terminologies being cut off from the world, one is led to consider their dissemination and thus the contradiction behind scientific questions: there is no field of expertise without concern for propagation (from the laboratory to a company, a customer, or a consumer) (Gambier 1991, p. 10).

To think otherwise would be, as Delavigne puts it, to "ignore the important diversity and heterogeneity of audiences and to overlook the constant impurity of discourse" (Delavigne 2003, p. 81). Thus, like Jeanneret, we consider that scientific research is not opposed to popularization as pure knowledge on the one hand, and its rhetorical expression on the other, but rather as two mediations, regardless of how expert knowledge is communicated (Jeanneret 1994, p. 85).

2.3. Scientific media discourse

In this section, we shall move on to discourses about science that are not aimed at peers. This excludes the first level in Figures 2.1 and 2.2. Véron has characterized this form of "asymmetric" communication by the terms "endogenous trans-scientific communication" and "exogenous communication about science" (Véron 1997, pp. 30–31). In the first case, the speaker is a "scientist, producer of knowledge" and addresses a nonscientist (Véron 1997, pp. 30–31). The *Phytoma* magazine is representative of this first level of popularization in the field of agroecology[6].

In this work, we will mainly focus on the second ("exogenous") level of popularization, where the speaker is not a scientist but a journalist (potentially with a scientific background), addressing the lay citizen. The daily press, when dealing with scientific subjects, is part of this second category of popularization, as illustrated by the last level in Figure 2.1 and Figure 2.2.

Let us now investigate the specificities of the language used in exogenous discourses.

2.3.1. *Language and variation in the popularization process*

Given its objective, presenting science in the media involves a series of textual adaptations, such as reformulations, paraphrases or explanations. Semantic shift (leading to polysemy) and terminological variation (leading to synonymy) are obvious consequences of these processes. These phenomena have been attested in various fields: computer science (Meyer and Mackintosh 2000); nuclear physics (Delavigne 2003); DNA sequencing (Calsamiglia and Van Dijk 2004); genetically modified organisms or mad cow disease (Moirand 2004); globalization (Toledano and Candel 2006); energy and the environment (Dury 2008; Dury and Lervad 2008); technology and medicine (Gaudin and Alexandru 2006, Delavigne 2015); space technologies (Condamines and Picton 2014) and agroecology (Ledouble 2019).

Other linguistic phenomena show the interpenetration of the general and specialized lexicon in discourse. We can mention discursive hybridity, which attests to the permeability (or porosity) of the boundaries between ordinary words and specialized terms (Rakotonoelina 2014a). This notion may concern the genre under evaluation, such as online conferencing on different news websites (Rakotonoelina 2014b), or Internet users' comments in relation to current events (Calabrese 2014). Beyond genre, hybridity can also depend on the field of knowledge (e.g. ecology, in the case of e-conferencing), or arise from mixing of knowledge and opinions (in the case of users' comments). Attruia (2020) also demonstrates a phenomenon of "contamination" of non-expert lexical items within specialized discourses, the latter remaining however normalized, in order to avoid an arbitrary imbrication between discourses. The author illustrates this with the word *stowaway* (in reference to insects) in official European Union publications on biodiversity: "Others [*exotic*

6 Phytoma. *La santé des végétaux*. Groupe France Agricole [Online]. Available at: http://www.phytoma-ldv.com [Accessed 22 February 2022].

species] are introduced involuntarily; they are contaminants in goods [...] or 'stowaways' onboard ships or in equipment".

These interconnections between words and terms have an impact on the interpretation of such discourses, and we wish to explore the relation between the language used in exogenous discourses and popular knowledge. As Mortureux (1988, p. 145) has already pointed out, "the effectiveness of popular science is at the cost of a process of adaptation and of an arrangement of words, so as to allow for intercomprehension"[7].

We will investigate what is left of the (language of) specialization in media discourses. Do the latter preserve traces of specialized terminology (or variant forms)? Do they create new ones?

Different perspectives emerge in response to these questions, which also depend on the genres and mechanisms for popularization. In what might be called "secondary discourses" (exogenous communication), LSP may be detectable as a trace and/or virtually undetectable. Moirand (1995) specifies that in certain discourses, if it is "untraceable", it is "because the discourses, whatever their field of reference, simply use the potentialities of the linguistic system in use" (p. xviii). In some others, we find traces of specialized terminology, but which are reworked, reformulated and mixed with the lexical stock of an "average" press reader.

Thus, according to Delavigne, one of the challenges of the popularization of science is the discursive "dissolution" of technical jargon, but without excluding it completely: "Popular science discourses only use a small part of the terminology. The popularizer chooses to only keep a small part of it, thus only involving a specific readership" (Delavigne 2005, p. 193).

The implicit question is then: what happens to specialization, if terminology is reworked or reformulated? As Moirand (2014) rightly argued, the question of specialization deserves more investigation, within discourses whose initial objective is not to convey specialized knowledge.

However, even if the objective of popularization is not a priori didactic, it seems relevant to focus on the relations between the process of science popularization and the education process, which we shall discuss below.

---

7 Moirand (2007, p. 31) notes that technical terms are sometimes reserved for sidebars, which include definitions and interviews with scientists or experts. The author also shows that a term such as "manipulation", which is also present in general language, is sometimes used in a figurative (or pejorative) sense in secondary discourses, and not in the primary sense of the dictionaries in use (close to being changed).

2.3.2. *Parallel between popularization and education*

The similarities between popularization and education turn out to be quite fruitful when we look at the circulation of discourses devoted to science. Indeed, the expected simplification of media publications devoted to scientific information imposes a form of pedagogy (e.g. the perspective of popular science considered as a non-formal education opportunity). Moirand elaborates on the characteristics of these discourses "which often oscillate between 'showing' and 'informing' – 'informing' being likely to vary from simple information to education, in other words, 'making sure that the reader knows'" (Moirand 2004, p. 79). Delavigne extends this parallel between the didactic discourse and the media discourse: "the popularizer, like the teacher, must act as a medium, an intermediary, a negotiator, a mediator, grappling with pre-existing conceptions that most often need to be modified [...], without expecting any necessary prerequisites" (Delavigne 2005, pp. 190–193). Charaudeau also associates them, specifying a point of convergence between the two that is their finality (at least partially): sharing "the dual purpose of informing (to make something known) and captivating (to arouse interest)" (Charaudeau 2008, p. 18).

The press would then represent some form of education, "when anecdotal elements are being generalized, or when encyclopedic knowledge is mingled and sometimes entangled with more expert knowledge" (Moirand et al. 1995, p. xiv).

Thus, even in the context of "ordinary discourse" such as that of the press, our initial hypothesis remains valid. We consider that we cannot properly inform the reader about a scientific subject without using a minimum number of elements of knowledge.

Various research works have been studying these questions for a long time, notably in the educational sector of English for Specific Purposes[8].

We refer, for instance, to some of Van der Yeught's studies, which explore these issues from both a theoretical and practical point of view, based on the following questions: what is the relationship between language and specialization? What is specialization? In terms of teaching, the author examines the specialized skills of the LSP teacher in relation to the LGP teacher, as well as to the specialists themselves:

---

8 See in particular the publications of study groups such as GERAS (*Groupe d'études et de recherches en anglais de spécialité*), through its *AsP* journal, GERES (*Groupe d'études et de recherches en espagnol de spécialité*) and the *Cahiers du GERES*, or the *Recherche et pratiques pédagogiques en langues de spécialité* journal – *Cahiers de l'APLIUT* of the *Association des professeurs de langue des instituts universitaires de technologie*.

> There is a question that has long preoccupied LSP practitioners: do they need to have specialized knowledge of the fields for the language in which they teach [...]? Should they limit themselves to being language teachers with no specialized knowledge or should they invest time and effort to master some of it? (Van der Yeught 2019, pp. 80–81).

We will now further explore the links between specialization and the educational approach within LSP, as they present relationships that seem as discrete as they are obvious, and are useful for furthering the question of the popularization of science.

2.3.2.1. *Specialization and didactics*

Petit attests to the social nature of any field of expertise, which he sees as a "sector of society built around and for the practice of a specific activity" (Petit 2010, p. 21). While Van der Yeught (2019) admits that "what is specialized" is eminently social, he intends to specify what is meant by "specialization". According to the author, to become specialized is to focus on notions that are specific to a social activity in connection with a profession or an area of expertise. Professional communities then generate a language for specific purposes to produce new sociolinguistic realities (listed in dictionaries and encyclopedias). Therefore, to better understand these innovations, does an LSP teacher or a journalist writing popular science need a minimum of "specialization"? Probably not, if we consider that their role is not to share knowledge or to train experts in the profession or field in question.

How, then, can we transmit lexical elements to non-experts, without being specialized in the professional domain to which the terms refer[9]? If we follow this reasoning, in the absence of LSP (in education or popular science), teachers or mediators could communicate, but not thoroughly explore or explain the objects, the processes, or even the issues related to a field of expertise.

How then can we achieve the paradoxical goal of teaching LSP or informing readers about a professional domain, without using the terms that convey the elements of specialization, which are the key to accessing knowledge?

---

9 In order to make this observation, the author notably relies on the field of English for Specific Purposes (ESP), in which he regrets the marginalization of the specialty, which is treated in a circumstantial logic. He illustrates his point through the variation of adjectival forms such as using English for medical purposes, for legal purposes, for academic purposes, where language is used as a communication medium, but where the areas of expertise that come into play are not conceptualized in an analytical way.

A relatively pessimistic perspective would substantiate the first critics of popular science, considering that, in order to popularize, we must "share knowledge", and the ambition of popularization would then seem pointless.

And yet, educators continue to educate and journalists continue to inform (on specialized subjects). Is this the transmission of minimal but superficial knowledge? Is it a futile process? This perspective does not seem to satisfy us.

In his intentional approach to specialized languages, Van der Yeught (2019) advocates a "third way", which is to no longer systematically oppose linguistic knowledge and specialized knowledge, but to put the area of expertise at the heart of LSP teaching. We will explore this question further when exploring science in the media. To this end, let us briefly address the subject of journalist training.

2.3.2.2. *Specialization and journalist training*

Some university courses prepare mediators to specialize in scientific fields[10]. Science journalists would thus ensure both the content and the editorial quality of the information transmitted. However, in view of the competition between the media and questions of employability, the versatility of a journalist is often favored. Peynaud has conducted a study on the training schemes for journalists, which shows that

> all of these schemes are not aimed at the acquisition of expertise in a specific field by students, but rather at the construction of a fairly broad general culture, which should enable them to deal with any type of subject involving current affairs (Peynaud 2014, p. 25).

The author adds: "the journalism courses that have been analyzed therefore aim at the acquisition of broad multidisciplinary skills by journalists, that are based on a shallow knowledge of a wide variety of specialized fields" (Peynaud 2014, p. 25). This principle is in line with Ruellan's analysis of the very foundation of the profession, who considers the "enriching and radiant" place of the journalist who is inspired by different genres (philosophy, education, scientific research, politics, literature, etc.) "without having to suffer from the potential limits imposed by specialization" (Ruellan 1992, p. 36). We nevertheless wonder about the legitimacy or the capacity of the mediator to explain subjects that they do not specialize in. Some recall the asymmetry principle, in which journalists will have acquired a

10 For example, the Communication and Scientific and Technical Culture (CCST) course, in the Master's degree course in Information-Communication at the Université Grenoble-Alpes, the "Scientific Journalism" Master's degree at the Université Paris Cité, the "Science Mediation" Master's degree at the Université de Bordeaux, or the Eco-citizen Communication, Heritage and Sustainable Development (COMEDD) Master's degree at the Université Côte d'Azur.

certain amount of knowledge about the subject matter (by reading press releases, interviewing a specialist, etc.), thus establishing themselves "more or less explicitly as possessing more knowledge than the person whom [they are] addressing" (Véron 1997, p. 31). They will also have been media-trained, and as such, able to encompass scientific information in a broader societal context, as described in section 1.2.1. They will therefore be particularly skilled in bringing up (and debating on) a subject in a multidisciplinary and global manner, in relation to its scientific, socio-economic, political and even ethical issues.

However, the situation may be different depending on the fields considered. According to Peynaud (2016), the media coverage of the Fukushima nuclear accident involved a relatively common phenomenon (an earthquake) that does not necessarily require specific technical or scientific knowledge. The author recalled the view of Sager et al. (1980)[11], which states that a specialized discourse must be produced by specialists in a field. Thus,

> although some authors have specializations in science, they are journalists first, before being characterized by any other type of specialization, [...], as a specialized discourse is supposed to be produced by specialists of the field (Peynaud 2016, p. 21).

In the field of nuclear accidents, Peynaud only identifies a minimum of specialized terminology (for questions of legitimization), but not scientific knowledge as such. However, these reflections are based on phenomena that are relatively well known to the general public (earthquakes, tsunamis, etc.). We then wonder whether this reflection also applies to agroecology, a field that is still not well publicized or known to the general public. The situation seems a priori different, as the processes and concepts involved are not familiar or "experienced" by the general public. This is precisely the question that concerns us here, which we will attempt to answer in Part 2 of this work.

Before moving on to the analysis of textual data in a media context, we shall present the manner in which we will analyze this textual data in Chapter 3.

11 Sager, J.C. et al. (1980). *English Special Languages. Principles and Practice in Science and Technology*. Brandstetter Verlag, Wiesbaden.

3

Theoretical and Methodological Tools for Analysis

Our study falls within the framework of textual data analysis and corpus linguistics[1]. By working on specialized discourses, the principles of a textual and descriptive approach to terminology will be followed (see Chapter 2). However, let us first address the methods used for the identification of names and designations within the corpus.

3.1. Identification and categorization of names and designations

The notion of "naming" has been broadly defined by Kleiber (1984, p. 77) as a "stable and recurrent relation between a linguistic expression and an extra-linguistic entity". For example, in our corpus, the English term *biological control* (in French *lutte biologique*) has historically been preferred as the main name of the concept we are interested in (and dates back to the 19th century). However, societal interest in these issues has generated other names for it in media discourse. We thus observe a list of simple (or complex) terms that form part of a "designation paradigm"[2]: "As a first approximation, these are lists of phrases (generally nominal, sometimes verbal) that function in co-reference with an initial term in a given discourse" (Mortureux 1993, p. 124). We therefore search for all of the names and designations in the press corpus, according to the methodology presented below.

1 The TXM platform (Heiden et al. 2010) was used for all corpus research.

2 From the original French term "*paradigme désignationnel*".

3.1.1. Designations in a discursive context

In order to identify these competitive designations (L'Homme 2004; Holzem and Labiche 2017), we start with the most common term and apply various "lexico-semantic tests", so as to identify competitive terms and "confirm a domain-specific meaning" (L'Homme 2004, p. 70). This method, known as the "substitution method" (L'Homme 2004, p. 95), relies on two important criteria that are "compatible co-occurrence" and "differential co-occurrence" (Mel'čuk et al. 1995, pp. 64–68), as well as substitution via a synonym.

The next paragraph illustrates the method with the term *biological control*:

– On the one hand, from the following pattern: "adjective + *control*" or "*biological* + noun", we identify new units that can be hypothetically substituted with *control* and/or *biological* (e.g. *biological regulation*, *biotechnical control*, etc.).

– On the other hand, we observe the contexts in which *biological control* (and the previously identified terms) appear in order to identify new linguistic forms, based on the two discursive axes (syntagmatic, which is the "horizontal" sentence axis, and paradigmatic, which is the invisible "vertical" axis with potentially lexical choices). New terms emerge, referring to the same extra-linguistic entity, which would not necessarily be indexed in a reference work, as shown in Table 3.1; we see substitutable units for *control* (*regulation*, in A), for *biological* (*biotechnical*, in B) and for the whole phrase "biological control" (*bioprotection*, in C).

	Simple and complex terms	
A	Biological	Control Regulation
B	Biological Biotechnical	Control
C	Biological control Bioprotection	

Table 3.1. *Denominative variation from the term "biological control"*

However, some phrases identified by this method, which meet the potential substitution criteria, do not fit into the designation paradigm of *biological control*. The term *biological monitoring* (constructed on model A) is an example of this; it refers here to another concept, which nevertheless remains in the field of agronomy[3]. It is therefore essential to analyze the context of the term in the corpus in order to decide which terms belong to the designation paradigm, as well as those to be excluded from the set of designations competing with *biological control*.

This same substitution method is replicated on new potential terms, with a view to collecting the most complete designation paradigm, thus following the principles of synergy between hypothetical–deductive and inductive methods for designation analysis (Condamines et al. 2012). Once the set of co-referential phrases is identified (in French and English), a categorization of related terms is established.

3.1.2. *Analysis and categorization of designations*

Kageura (2002) presents a relevant method for classifying terms in a specialized field. In his study, the head term of a compound carries the main conceptual category, whilst the qualifying adjective, or modifier, subcategorizes it into potential co-hyponyms. For each main term of a phrase (e.g. *lutte* in French, or *control* in English), the associated modifiers (e.g. *lutte verte/bio, biological/organic control*, etc.) allow us to distinguish the set of co-hyponyms and their semantic specificities. This method "is formally equivalent to that of the description of semantic patterns of compounding general words" (Kageura 2002, p. 47), and can thus be effectively applied to the analysis of designations in news articles regarding a specialized field.

This methodological framework is then applied to the designation paradigm of plant protection methods and products, the results of which are presented in Chapter 5.

The analysis then extends to the lexico-discursive structures in which they appear, and to the characterization of these elements as given by the co-text and the context of the press articles.

3.2. Characterization of objects of discourse

The second part of the data analysis focuses on the link between textual structures and knowledge in the mediation process. Following the work of Moirand (2007), or Calsamiglia and Van Dijk (2004), our study is conducted around the three concepts of "text, context and knowledge, which articulate the discursive, social,

---

3 Biological monitoring (or biomonitoring) is "a series of processes and methods used to estimate the exposure of organisms and humans to toxicants by measuring their biomarkers in biological fluids and tissues." (source: https://www.sciencedirect.com/science/article/abs/pii/B9780128124666000026).

and cognitive aspects of popularization and of communication in general" (Calsamiglia and Van Dijk 2004, p. 385).

In order to examine these elements, we "move away from the notion of 'word' (and that of name or designation) for the more flexible notion of 'discourse object' [...]. One is then led to collect the expressions which reformulate, exemplify and enrich the representation of the different concepts within a corpus"[4] (Moirand 2007, p. 59).

Let us return to this notion of discourse objects, and to what we mean by the characterization of these objects.

3.2.1. *Characterization of an object of discourse: media schematization*

Grize developed the notion of discourse objects, along with the "natural logic" principles (1990, 1992). This seems relevant when we look at "science being explained"; objects are "constructed, proposed, negotiated, modified, refused or ratified in and by discursive processes" (Mondada 1995, p. 57), and "it is through the texts that they emerge contextually and are dynamically transformed" (Mondada 1995, p. 57). There is a relation between terms and discourse objects, and the latter enable readers to build their own linguistic and conceptual representation, and thus to acquire knowledge about the terms.

The objects of discourse at the heart of our study are the methods, agents and products of biological control; we are interested in the way they are presented in our French and English corpora – in other words, how they are cited, reworked and reformulated from a discursive perspective (Sitri 2003)[5].

According to Grize (2004), the way in which the journalist presents – and therefore characterizes – the objects of discourse can be described as "shedding a new light on some of their facets and concealing others". It is the process of "schematization" which gives a better representation of a situation[6] (Grize 2004).

4 Original French version: "*On abandonne alors la notion de mot (et celle de nomination, dénomination ou désignation) pour celle plus souple d'objet de discours [...]. On est alors conduit à relever les expressions qui reformulent, exemplifient, enrichissent la représentation de ce dont on parle*" (Moirand 2007, p. 59).

5 Sitri (2003) proposed a discursive analysis model of discourse objects based on this principle. This involves the object class, which includes the object itself and all its parts, or "ingredients", in the discourse.

6 Original French version: "*Donne à voir, mieux à regarder, une situation dans laquelle se trouvent des objets*" (Grize 2004, p. 42).

This relates to the process of “framing information”, and in particular “media framing“. According to Entman[7]:

> to frame is to select some aspects of a perceived reality and make them more salient in a communicating text, in such a way as to promote a particular problem definition, causal interpretation, moral evaluation, and/or treatment recommendation for the item described (Entman 1993, p. 52).

As a result, framing appears as a set of mechanisms that are more or less perceptible, and which lead the reader to infer their own value judgment[8] (Grize 1992, p. 25). They are thus involved in some form of argumentation, because “no one seriously questions his own conclusions”[9] (Grize 1992, p. 22). Nevertheless, if framing is the result of discursive choices made by the journalist, the intentional effects are moderated both by connotations involved (according to a dialogic perspective) and the reader’s interpretation. Varela already expressed the fact that “the communication act does not result in a transfer of information from the sender to the receiver, but rather in the mutual shaping of a world by means of a collaborative action” (Varela 1988, pp. 114–115). This perspective of “meaning co-construction” is shared by many disciplines (philosophy of language, psychology, sociology, etc.), and Charaudeau (2004, 2011) develops it in the context of a “communication contract”. He considers that the interpretation process on the recipient’s part belongs to social norms – and consequently to an axiological perspective of the world – corresponding to different points of view:

> ethical (what is good or bad), aesthetic (what is beautiful or ugly), hedonic (what is pleasant or unpleasant), pragmatic (what is useful or useless, effective or ineffective), which have become more or less stereotyped judgments that circulate in society (Charaudeau 2011, p. 34).

The author refers to connotated words (he cites “immigrant” and “illegal” in his study) and the “interpretive impact” that appeal to the reader’s evaluative response. He thus contrasts a body of expert knowledge, which does not require a specific evaluation, with another type of “believed-in knowledge”, established according to

---

7 And other studies on framing (e.g. Gamson and Modigliani 1989; Goffman 1991; Nisbet 2009).

8 Original French version: “*L’éclairage apparaît tel un résultat de mécanismes plus ou moins consciemment mis en place et qui conduisent le lecteur à inférer par lui-même un jugement de valeur*” (Grize 1992, p. 25).

9 Original French version: “*Personne ne met sérieusement en doute les conclusions qu’il a lui-même tirées*” (Grize 1992, p. 22).

cultural or social criteria, and potentially referring to a more subjective appreciation of information[10] (Charaudeau 2011, p. 34).

Understanding is therefore based on a social context (Van Dijk 1999), whose structure and schematic properties "organize and reduce the complexity of the social situation in such a way that language users have an efficient device to contextualize discourse production and comprehension" (Van Dijk 1999, p. 131). Knowledge is not always explicitly stated in media discourse, but it contains an implicit element: "It is this social nature of shared knowledge that defines presupposition and that allows discourse to be understandable without making all relevant knowledge explicit all the time" (Van Dijk 2003, p. 86).

In Chapter 6, different types of schematizations will be revealed, by delving into a particular form of framing, characterized by a binary polarity. This involves contrasting two elements and systematically showing their opposition within media articles. This relationship is part of what Bednarek and Caple (2014) describe as a form of "news values", referring to the classic duality between negativity and positivity, and thereby guaranteeing both the value and interest (newsworthiness) of information (among other works, see Nelkin (1986), Allan (2009), Bourassa et al. (2013), Labasse (2012) and Molek-Kozakowska (2017))

As Grize himself points out, one difficulty lies in the identification of the schema, and in the relationship between linguistic structures and the involved framing. It is now time to turn to the identification of the structures that give discourse objects a specific meaning.

3.2.2. *Formal markers characterizing objects of discourse*

Explanation, as a prototypical textual structure used in scientific mediation, does not necessarily rely on the classical argumentation markers (concession, objection, etc.), but on shorter textual structures that enable a quick and easy interpretation. We thus focus on the antonymy relation (see Cruse 1986; Jones 2002; Murphy et al. 2009), as it prefigures what we hypothesize, namely that media extensively relies on an opposition between methods and products of plant protection for the simplified presentation of discourse objects (as demonstrated for the English-language corpus in Ledouble (2020)).

The antonymy relation is considered in its broadest sense, characterizing "all types of lexical opposites" (Cruse 1986, p. 204). In his extensive study of antonymy,

---

10 The article elaborates on this distinction (see: https://e-cours.univ-lr.fr/UNT/mediatisation/res/texte-charaudeau02b.pdf [Accessed January 21, 2022]).

Jones (2002) identifies different discursive functions characterizing this semantic relation, four of which are relevant to our study: coordination, transition, comparison and negation. The author specifies that each functional category is based on a lexico-grammatical pattern, which allows the identification of relations in a corpus.

Our analysis starts from these different markers and their equivalents in French, searching for antonymous relations in our dual corpus, such as in the comparative study between English and Swedish presented by Murphy et al. (2009). These markers, as well as the four discourse categories of antonymy that shape our discourse objects, are briefly presented below.

Coordinated antonymy is a relation of opposition between two elements, but in which, paradoxically, the distinction between the two is neutralized "to signal inclusiveness or exhaustiveness of scale" (Jones 2002, p. 61). The markers that identify this relationship are presented in Table 3.2.

Comparative antonymy is a relation implying a measuring aspect, by comparing two elements on different criteria. It is usually indicated by the comparative markers shown in Table 3.3.

French	English
- *X et Y*	- X and Y
- *Aussi bien X que Y*	- Both X and Y
- *Tant X que Y*	- X and Y alike
- *Que ce soit X ou Y*	- Either X or Y
- *X en complément de Y*	- X as well as Y
- *X ou Y*	- X or Y
- *Qu'il(s) soi(en)t X ou Y*	- Whether X or Y

Table 3.2. *Markers of coordinated antonymy*

French	English
- *Plus X que Y*	- More X than Y
- *Moins X que Y*	- Less X than Y
- *Autant X que Y*	- X, no less than Y
- *Le plus/le moins*	- The most/the least

Table 3.3. *Markers of comparative antonymy*[11]

11 Some formulations fall under a sub-categorization of comparison, which Jones defines as the antonymy of distinction, introduced by the marker *between*.

Transition antonymy is a relation that implies a change, from one state or activity to another, which the markers listed in Table 3.4 are associated with.

French	English
- *X à la place de Y*	- X in place of Y
- *X au lieu de Y*	- From X to Y
- *Passer de X à Y*	- X gives way to Y
- *X céder la place à Y*	- X replaces Y
- *X remplacer Y*	- Turning X into Y

Table 3.4. *Markers of transition antonymy*

Lastly, negation antonymy refers to a co-occurrence between two objects, one of which is rejected and the other one "promoted". The markers that can be associated with this relation are listed in Table 3.5.

French	English
- *Pas Y mais X*	- X, not Y
- *Exit X, place à Y*	- not X but Y
- *Fini X*	
- *Plutôt X que Y*	- X instead of Y
- *X opposé à Y*	- X as opposed to Y

Table 3.5. *Markers of negation antonymy*

The set of theoretical and methodological tools presented in this chapter will allow us to analyze how specialized discourse objects are presented in Part 2 of the book.

Our questions in Part 2 will be the following: what is the competitive paradigm of terms in this field of expertise, both in English and French? Which elements of knowledge are conveyed through discourse objects? What representation, or co-construction, of knowledge emerge from these publications? Can a citizen engage in an informed debate on these subjects, based on media discourse?

Our analytical (and non-critical) approach will attempt to examine a reader's potential perception of the "social explanation" of a highly complex field, such as agroecology.

PART 2

Science in the Media: Agroecology in the Daily Press

Introduction to Part 2

This part introduces a set of media strategies, already attested in previous research work (Ledouble 2019, 2020, 2021), but presented here according to a double perspective: on the one hand, the issues of naming methods, agents and products used for plant protection using natural mechanisms; on the other hand, their contextual description, hypothesizing a specific framing. In this context, the linguist has two roles: they first place themselves in the reader's position (as a non-specialist) to identify how the latter perceives specialized information via press articles; they also examine the ambiguous terminology within these publications. Our study is based on a number of corpus examples, so as to illustrate all analyses with the provided data.

Before presenting the results of the textual analysis in Chapter 5 and Chapter 6, we shall first present the field at the heart of this work, as well as the corpus used for this case study.

4

An Introduction to Agroecology

A truly extraordinary variety of alternatives to the chemical control of insects is available. Some are already in use and have achieved brilliant success. Others are in the stage of laboratory testing. Still others are little more than ideas in the minds of imaginative scientists, waiting for the opportunity to put them to the test.

Rachel Carson (1962, p. 242)

In the field of crop protection, the use of "alternative" methods to chemical pesticides has been encouraged by recent regulations[1]. These legal initiatives have led to greater media exposure, which have been encouraged since 2017 by certain legislative shifts concerning glyphosate and neonicotinoid insecticides (Allard-Huver 2021) (from their prohibition to their reintroduction under certain conditions of use, of time, or for certain crops)[2].

Agroecology is closely linked to these regulations, with natural methods being proposed as an alternative to these controversial products. However, despite increased media coverage, biological control still seems to be relatively unknown to the general public at the time of writing. Surveys conducted among non-specialists

---

1 For example, the European Proposal for a regulation in June 2022, for the sustainable use of pesticides (see: https://ec.europa.eu/info/law/better-regulation/have-your-say/initiatives/12413-Pesticides-sustainable-use-updated-EU-rules-_en), or two regulations in 2014 to better regulate the use of plant protection products on the French national territory (see: https://www.legifrance.gouv.fr/eli/loi/2014/2/6/2014-110/jo/texte, and https://www.legifrance.gouv.fr/eli/loi/2014/10/13/2014-1170/jo/text) [Accessed April 2023].

2 See: https://www.assemblee-nationale.fr/dyn/15/rapports/cion-eco/l15b1677_rapport-fond [Accessed February 15, 2022].

show that the main (if not the only) reference to this field is the ladybird, which feeds on aphids and thus protects the plant from potential pests. Although limited, it is also an efficient reference to the principles of biological control, and the press, as we shall see, uses it thoroughly (as does the industrial and advertising sector, for whom the ladybird remains an iconic representation of this field). However, this is only one of the many processes involved in these alternative methods.

Faced with a complex innovative subject that is somewhat obscure to the lay citizen, the media has a major role: to explain both the objectives of the field (and its innovative methods), as well as the associated societal issues.

Before presenting the attested denominative variation in Chapter 5, it is important to recall the terminological elements from a scientific and legal perspective, as well as the equivalents between French and English terms.

4.1. Presentation of biological control

The generic term for all (complementary) methods of plant protection is *Integrated Pest Management* (IPM). The Directive of the European Parliament and the Council of October 21, 2009 proposes the following definition:

> Integrated pest management means careful consideration of all available plant protection methods and subsequent integration of appropriate measures that discourage the development of populations of harmful organisms and keep the use of plant protection products and other forms of intervention to levels that are economically and ecologically justified and reduce or minimize risks to human health and the environment (European Commission – Food Safety)[3].

The objective of this directive is to establish a framework for community action, so that pesticide use is compatible with sustainable development. A seminal scientific article on these issues (Eilenberg et al. 2001) provides a relevant diagram for understanding the structuration of IPM, which is useful for our study.

In Figure 4.1, we have highlighted the two essential elements in our study, biological control and natural substances of vegetal or mineral origin, which we will define in the next section.

3 Source: https://food.ec.europa.eu/plants/pesticides/sustainable-use-pesticides/integrated-pest-management-ipm_en [Accessed January 2024].

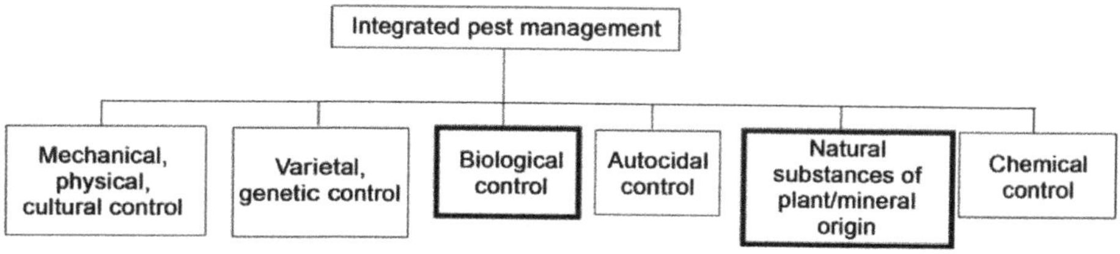

Figure 4.1. *Reproduction and adaptation of Figure 1 (from Eilenberg et al. 2001)*

4.1.1. *Definition of terms and terminological instability*

Biological control is defined as "the use of living organisms to suppress the population density or impact of a specific pest organism, making it less abundant or less damaging than it would otherwise be" (Eilenberg et al. 2001, p. 390). In this approach, biological control agents are living organisms (macro-organisms such as insects, invertebrates, mites, nematodes or micro-organisms such as bacteria, fungi or viruses).

Natural substances of plant or mineral origin (equivalent to bio-chemicals, botanical pesticides or biopesticides) correspond to substances naturally present in a vegetal environment (plant extracts, essential oils, etc.) or a mineral one (potassium, kaolin, copper, sulfur, etc.).

These plant protection products and methods are therefore part of the generic integrated pest management approach, along with the other types of control(s) represented in Figure 4.1 (mechanical, physical, cultural, varietal, genetic, autocidal and chemical control). We wish to emphasize the fact that chemical control is also part of integrated pest management; however, its final position on the diagram indicates that it should only be used as a last resort (and "at reasonable levels", according to the European directive of 2009) when other control methods fail to regulate the flow of pests. It is also important to note that the fundamental IPM principle is not to exterminate pests (which is the case with chemical control alone), but to "discourage" them (Eilenberg et al. 2001, p. 389), therefore limiting and regulating their numbers. Conventional, wide-range chemical treatment is generally aimed at eradicating pests, and therefore does not theoretically meet this "regulation" principle.

This area of specialization is based on several scientific disciplines (agronomy, entomology, plant pathology, biology, chemistry, etc.), which has led to a linguistic instability that was already denounced by Eilenberg et al. in 2001, arguing that "the different traditions in the disciplines of biological control workers have, however, led to the lack of a common terminology" (Eilenberg et al. 2001, p. 388).

Over 20 years later, the teminological situation is still problematic, particularly with regard to the names, designations and status (whether living or not) of these products and methods. It becomes even more complex when we consider the issue from a cross-linguistic point of view, particularly since the enactment of the Law for the Future for Agriculture, Food and Forestry (LAAF) in 2014: it introduces the French term *biocontrôle*, which was not previously used, forged from the English term *biocontrol*.

The legal text defines the term *biocontrôle* as follows: all "agents and products using natural mechanisms within the framework of integrated pest management". The law thus modifies Article L 253-6 of the Rural and Maritime Fishing Code (CRPM), which specifies that these agents and products "include the following:

1) macro-organisms;

2) plant protection products comprising microorganisms, chemical transmitters such as pheromones and kairomones, and natural substances of plant, animal or mineral origin"[4].

These texts thus impose the French term *biocontrôle* as the official name for these products and methods, in other words, a term that "falls within the legal framework of a legitimized institution and operates within a perspective of standardization of naming procedures" (Petit 2012, p. 39). It is opposed to a "de facto name" which, according to the same author, "emerges from collective usage" (Petit 2012, p. 39). Three issues arouse from this official definition.

1) The first one is in the choice of the term *biocontrôle* in French, based on the English term *biocontrol*, despite the fact they are not equivalent. *Biocontrol* is well established in international communication and corresponds to the French term *lutte biologique* (which refers to living organisms). The term *biocontrôle* includes a list of products that are not only living organisms but also natural substances of plant, animal and mineral origin. Biological control is therefore only a small part of biocontrol (Deguine and Ledouble 2022).

A second issue emerges from this first point, relating to the categorization of biocontrol products.

2) Micro-organisms, which are (living) biological control agents, are categorized differently in the French regulation, which presents them as "phytosanitary products"[5].

4 Law No. 2014-1170 of October 13, 2014, on the future of agriculture, food and forestry, 50(1) [Online]. Available at: https://www.legifrance.gouv.fr/eli/loi/2014/10/13/2014-1170/jo/article_50 [Accessed 15 January 2024].

A third issue emerges from this name, which will prove to be important in our study.

3) The choice of the term *phytosanitary product* as a generic term to designate biocontrol products raises debatable questions in terms of communication to nonspecialist readers. Indeed, this term is generally associated with chemically synthesized products (or pesticides) and has a depreciatory connotation in the eyes of the layperson. However, in this case, it refers to biocontrol products, which reproduce "natural mechanisms" for plant protection[6].

These differences in terminology and categorization present a challenge in terms of communication, especially to a broad and heterogeneous readership.

Let us briefly discuss the international legal approach to this area of specialization.

4.1.2. *Regulatory approach to plant protection*

The European and U.S. regulations have their own specificities. The European Union did not publish any official definition for biocontrol until June 2022, which caused a number of difficulties[7]. In a recent bill, the European Commission introduced and defined the English term *biological control*, the French equivalent being *protection biologique* (and not *lutte biologique*)[8]. This European definition is close to the French definition of *biocontrôle*, although it does exclude mineral products (Deguine and Ledouble 2022). This partial terminological correspondence between French and European laws is likely to generate misunderstandings on an international level. Furthermore, if we refer to the European Commission's official website[9], the generic English term *plant protection*

5 In French, *produits phytosanitaires*, which is equivalent to "plant protection products", but has a specific connotation (i.e. issue No. 3, below).

6 It is emphasized here that biocontrol (plant protection) products also rely on chemical processes (to mimic natural processes), so biocontrol should not be considered as chemical-free.

7 As stated in the proceedings of a "Question/Answer" session in the French Senate, "this legal gap makes the system complex and causes interpretative issues between Member States. The application of the regulation is consequently unclear and affects evaluations and market authorizations, as well as the control system". See: https://www.senat.fr/questions/base/2018/qSEQ18060377S.html (JO Senate, June 7, 2018) [Accessed 24 February 2022].

8 Regulation of the European Parliament on the sustainable use of plant protection products, and amending Regulation (EU) 2021/2115 [Online]. Available at: https://ec.europa.eu/info/law/better-regulation/have-your-say/initiatives/12413-Pesticides-sustainable-use-EU-rule-actualized-_en [Accessed April 2023].

9 See: https://ec.europa.eu/food/plant/pesticides_en (see the multi-language view).

product (defined in the European regulation)[10] is equivalent to the French term *phytosanitary product* (and to the term *pesticide*). Their simplified definition appears on the EU's official page:

> Plant protection products are "pesticides" that protect crops or desirable or useful plants. [...] They contain at least one active substance [which] is any chemical, plant extract, pheromone or micro-organism (including viruses).

The terms *plant protection product* and *pesticide* thus encompass all plant protection products and methods (both natural and chemical).

The Environmental Protection Agency in the United States (EPA) favors the term *biopesticide* (or *biological pesticides*) with the following expanded definition:

> Biopesticides include naturally occurring substances that control pests (biochemical pesticides), microorganisms that control pests (microbial pesticides), and pesticidal substances produced by plants containing added genetic material (plant-incorporated protectants) or PIPs (1998)[11].

International researchers do not necessarily agree on what the term *biopesticide* means from a scientific perspective. Shortly after this regulation, Eilenberg et al. stated that "there are strong arguments in favor of reserving the term *biopesticide* for preparations containing living (micro-)organisms" (Eilenberg et al. 2001, p. 388).

This approach was contradicted a few years later by American researchers Pal and McSpadden Gardener (2006), who stated that although these products "may mimic the activities of living organisms, non-living inputs should more properly be referred to as *biopesticides*".

Furthermore, the FAO (Food and Agriculture Organization of the United Nations) defines *biopesticides* according to their usage, associating them with chemical synthesis pesticides:

> A generic term, not specifically definable, but generally applied to a biological control agent, usually a pathogen, formulated and applied in a manner similar to a chemical pesticide (FAO 2006).

10 Regulation (EC) No. 1107/2009 of the European Parliament of October 21, 2009, concerning the market of plant protection products 79/117/EEC and 91/414/EEC (JO L 309, 24/11/2009, p. 1) [Online]. Available at: https://eur-lex.europa.eu/legal-content/FR-EN/TXT/?uri=CELEX:02009R1107-20181110.

11 See: https://www.epa.gov/pesticides/biopesticides [Accessed on January 18, 2024].

The OECD offers yet another definition for *biopesticides*, which includes both micro- and macro-organisms, as well as pheromone-like substances and plant extracts (Food Standards Programme Codex Committee on Pesticide Residues 2018)[12].

These different scientific, institutional and legislative perspectives are likely to generate confusion in terms of communication (Eilenberg et al. 2001; Pal and McSpadden Gardener 2006).

This complexity is still present nowadays, as confirmed by Bernard who highlights a certain "fluctuation" in the definition of *biological control* and the associated terms, based on the authors' perspectives or specializations:

> Sometimes the definition includes genetically modified plants or microorganisms, which others reject. Many terms vary depending on the nature of the pests to be eliminated, the origin of the biocontrol agents studied, the time of human intervention in the crop cycle, etc.[13] (Bernard 2017, p. 9).

Faced with an eminently complex terminological and cognitive situation, our goal is to analyze how the media covers these subjects and, incidentally, press readers' perceptions of them. We have built a double corpus in an attempt to investigate this question, which we present in section 4.2.

4.2. Media discourse under study

The field of biological control (and biocontrol) does not (yet) have significant media coverage, even though it has been increasing every year for the past decade (Ledouble 2019). Without knowing how it will evolve, we have chosen to explore press articles devoted to it since the beginning of its media coverage in French and English daily newspapers (starting in 2010). Collecting and analyzing articles over a short and recent diachrony is an approach that is becoming more common in textual terminology (Picton 2018), and constitutes "a fundamental methodological option

---

12 See: http://www.fao.org/fao-who-codexalimentarius/sh-proxy/en/?lnk=1&url=https%253A%252F%252Fworkspace.fao.org%252Fsites%252Fcodex%252FMeetings%252FCX-718-50%252FCRDs%252Fpr50_crd03e.pdf.

13 Original French version: "*Parfois la définition englobe des plantes ou des micro-organismes génétiquement modifiés, ce que d'autres rejettent. Beaucoup de présentations varient selon la nature des nuisibles à éliminer, l'origine des agents de biocontrôle étudiés, l'époque de l'intervention humaine dans le cycle des cultures, etc.*" (Bernard 2017, p. 9).

for attesting to the multiform heterogeneity of the discursive units"[14] (Moirand 2007, p. 10). We address all collected articles as a whole, regardless of their form (editorials or standard articles), while theoretically considering it unlikely that the average reader would link them to the genre of the article (Moirand 2007, p. 22). Our objective, as analysts of textual data, is to capture and even "model" the perception of the field (from both a macro and a micro perspective) that a non-specialist press reader may have when exposed to this thematically homogeneous global corpus.

The growing public interest to alternative methods invites us to focus on a short, recent diachrony (Dury 2006; Tolédano and Candel 2006; Humbley 2011; Picton 2018). Our corpus consists of news articles published between January 1, 2010, and December 31, 2019: a period of 10 years before and after the regulations published in 2014 and which are important to our study.

We therefore have a double corpus for the analysis of textual data, both in French and in English. Our objective is not to carry out a precise comparative analysis between French and English (as the corpora is not identical on all points). Nevertheless, because of the influence of the English terminology on the French one, we have opted for an analysis on two independent corpora, which present similarities: they both deal with the same area of specialization covered by the daily press, with both corpora being collected on the same timescale.

4.2.1. *French press corpus*

The French corpus is built on national and regional daily press articles. Their selection was based on the three keywords: *lutte biologique*, *protection biologique* and *biocontrôle*. These terms were recommended by experts for their capacity to gather an exhaustive set of articles that discuss alternative practices for plant protection by natural mechanisms. The collected corpus includes all press articles published in the selected time span, including at least one of the three aforementioned keywords, from 14 reference newspapers of the French general press[15].

---

14 Original French version: "*[Cet approche constitue] une option méthodologique fondamentale pour attester de l'hétérogénéité multiforme des unités discursives recueillies*" (Moirand 2007, p. 10).

15 The selected titles were among the largest daily press circulations according to the ACPM ranking on the date the corpus was compiled (source: 2022). The Europresse and Factiva press databases were used to collect the articles in the corpus. In terms of keywords, we also considered graphic variations such as *biocontrôle*: *bio-contrôle* and *bio contrôle*.

It is composed of sources listed in alphabetical order, with the number of corresponding articles in parentheses:

– French regional daily press: *La Montagne* (152), *La Nouvelle République du Centre* (148), *La Voix du Nord* (79), *L'Est Républicain* (82), *Le Parisien* (36), *Le Progrès* (139), *Ouest-France* (516).

– French national daily press: *La Croix* (44), *La Tribune* (37), *Le Figaro* (53), *Les Échos* (38), *L'Humanité* (23), *Le Monde* (40), *Libération* (16).

The French-language corpus is thus composed of 1,403 press articles.

4.2.2. *English-language press corpus*

The English-language corpus is based on the national press in British and American newspapers. The articles were selected using the following keywords: *biological control*, *biocontrol*, *biopesticide(s)* and *biological pest control*[16]. The corpus thus includes all articles published between 2010 and 2019, containing at least one of the above-mentioned keywords, from 11 British and American newspapers.

The corpus includes the following sources, with the corresponding number of articles in parentheses:

– British press: *Financial Times* (12), *The Guardian* (46), *The Independent* (15), *The Daily Mail* (5), *The Daily Telegraph* (69), *The Sun* (71), *The Times* (24).

– American press: *The New York Times* (11), *USA Today* (4), *The Washington Post* (16), *The Wall Street Journal* (10).

The corpus consists of 283 national press articles.

The two corpora therefore have common points (daily press discourse, short diachrony, keywords), but the English-language one does not include the regional press. Through textual data analysis, our aim is to describe the terminological instability in this field, in the two languages of study and in the selected corpora.

We now move on to the presentation of the results of this analysis, starting with naming and designation issues.

16 These keywords were also recommended by experts to cover related issues.

5

Names and Designations of Discourse Objects

This chapter will focus on the different names and designations of plant protection methods, agents and products in the double corpus, by applying the methodological framework defined in Chapter 3. We will first present the simple and complex terms that are equivalent to the French compound *lutte biologique* and to the English term *biological control* in media discourse, before giving a summary of these names and designations in both languages at the end of each section.

5.1. Designations of plant protection methods

We will start with the terms referring to protection methods in the French corpus.

5.1.1. *Protection methods in French*

The term *lutte biologique (biological control)* is the original French denomination for the domain. The following is a corpus extract illustrating the use of this term in press discourse:

> 1) "La Société horticole de Sens propose de découvrir les méthodes de *lutte biologique* contre les parasites et ravageurs des cultures"[1] (*La Montagne*, 2011).

Another compound is regularly used to refer to the same practices, substituting the word *lutte* with *protection*:

---

1 English translation: "The Horticultural Society of Sens suggests discovering the methods of biological control of crop parasites and pests" (*La Montagne*, 2011).

> 2) "À la place de pesticides, la ville a développé une *protection biologique* grâce à la présence d'un insecte"[2] (*Ouest-France*, 2011).

The two elements of the compound are important to understand the notions involved (*protection* + *biologique*). This terminological choice in specialized discourse usually depends on the perspective: insect specialists will tend to favor the use of the term *lutte (control)*, whereas plant specialists prefer the term *protection*, even if the final objective is similar and the methods are interdisciplinary (pest *control* for plant *protection*). This recalls the "multidimensionality of concepts" (see Chapter 2), which is part of the natural and functional language capacity.

However, in the study of press discourse, it is not always easy to identify the reasons for this terminological choice and, therefore, the cause of the naming variation (which can also be explained by narrative or framing elements, to which we shall return). At this stage, if we follow the classification of denominative variation proposed by Freixa (2006), the creation or use of a variant can be related to a discursive strategy (e.g. favoring non-redundancy); however, it can also be a cognitive type of variation, which depends on the way the elements of the field are conceptualized.

In addition to these two classical predicates (*lutte* and *protection*), press discourses substitute other forms, such as the lexical unit *contrôle*:

> 3) "Koppert s'est fait une spécialité du *contrôle biologique* des prédateurs d'arbres"[3] (*Les Echos*, 2016).

We can hypothesize that this is a matter of interlinguistic variation (originating from contacts between languages), if we refer to Freixa's classification (2006); indeed, the English term *biological control* (or *biocontrol*) is the main name for this field in English, and the French term *contrôle biologique* is a loan translation of the latter. French-speaking discourse often tends to come close to the English terminology (by borrowing or reproducing a literal structure), as a form of knowledge standardization. In this case, the term *contrôle biologique* is used in French, despite the recommendations of certain institutions[4].

---

2 English translation: "Instead of pesticides, the city has developed *biological protection* thanks to the presence of an insect" (*Ouest-France*, 2011).

3 English translation: "Koppert has specialized in the *biological control* of tree predators" (*Les Échos*, 2016).

4 The English-language meaning of *control* refers to the regulation of pests. The French language, by semantic proximity, uses the term "*contrôle biologique*" with this same meaning, but inappropriately. The Terminology Commission, which has established the

We observe that the three predicates (*lutte*, *protection* and *contrôle*) are associated with the term *biologique*. The reference definition of the adjective *biologique* (or *bio*) refers to biology and, as a consequence, to life and living organisms.

These four terms are built on lexical units that are also used in ordinary language (*biologique*, *lutte*, *protection*, *contrôle*). With no specialized definition, the reader is likely to rely on the lay meaning of each lexical unit (and their hypothetical compositional meaning), to interpret the term in context (see Chapter 7 for the impact of these terminological choices on the perception that a reader may have of the domain or methods involved).

The term *biocontrôle* emerges in press articles after the publication of the French Law for the future of Agriculture, Food and Forestry in 2014:

> 4) "Le plan prévoit de favoriser […] les techniques de *biocontrôle*, c'est-à-dire la protection des végétaux par des mécanismes naturels"[5] (*La Croix*, 2015).

The press also relies on forms derived from this neologism, such as the unit *bioprotection*, which is formed on the semantic model of the term *biocontrôle*:

> 5) "Cette technique de '*bioprotection*', qui vise à la protection des végétaux par des mécanismes naturels […]"[6] (*Le Monde*, 2014).

The single quotation marks around *bioprotection* show a certain "distance" towards this term, with the journalist potentially indicating that they are "not able to take responsibility for the innovation on his own account, and presenting himself only as a transmitter"[7] (Pruvost and Sablayrolles 2003, p. 69). This common technique in media discourse is generally applied to a potentially unconventional or ambiguous unit. Another example in our corpus is the use of the qualifier *bio* (within quotation marks) to describe the control process:

---

terminological glossary of the French Association for Plant Protection (AFPP) since 2007, declares that "the French term *contrôler* is a misnomer, based on the English word *to control*. The French term 'maîtriser' would be more appropriate, semantically speaking" (Source: Répertoire terminologique en protection des plantes, 2007, 2011, 2017).

5 English translation: "The plan is to promote […] *biocontrol* techniques, in other words, plant protection by natural mechanisms (*La Croix*, 2015).

6 English translation: "This '*bioprotection*' technique, which aims to protect plants through natural mechanisms […]" (*Le Monde*, 2014).

7 Original French version: "Le fait de ne pas assumer pour son propre compte la responsabilité de l'innovation et de ne se présenter que comme un transmetteur" (Pruvost and Sablayrolles 2003, p. 69).

> 6) "La *lutte 'bio'* prend racine aux espaces verts municipaux"[8] (*Le Progrès*, 2016).

The use of "bio" (instead of "biologique") raises questions as to the specialization of the term. Indeed, the adjective *biologique* (or its abbreviated form *bio*) has a second meaning in day-to-day French language, which is "without pesticides or synthetic fertilizers". As a result, two interpretations are possible, whether the newly forged collocations are based on the first or second meaning of the term (or both).

For example, a "*lutte 'bio'*" may refer to a method using living beings (which is the first meaning of *bio*), or one that does not include chemical synthesis products (second meaning). This ambiguity becomes all the more problematic as other equally polysemous terms complete the designation paradigm. Thus, a "*lutte verte*" (*green control*) is equivalent to *biological control* in discourse:

> 7) "Une nouvelle stratégie de *lutte verte* mise en œuvre dès cette année pour soigner les platanes de Menton"[9] (*Ouest-France*, 2012).

In this sentence, the adjective *verte* (*green*) specifies the control method. However, from a semantic point of view, it refers to the impact of these methods on the environment. Other synonymous forms have been identified in the corpus, such as the unit *ecologique (ecological)*:

> 8) "[...] La *lutte* biologique et *écologique*, un budget stable, voilà quelques règles que nous nous sommes imposées"[10] (*Le Progrès*, 2013).

We note the emergence of a lexical field around the notion of environmental protection. Thus, as Mortureux points out, "the (*specialized*) term does not necessarily appear – in spite of its precision, its appropriation – as the most accurate in all circumstances, because some of its reformulations may have the advantage of

---

8 English translation: "The *'bio'* control takes root at municipal green areas" (*Le Progrès*, 2016).

9 English translation: "A new *green control* strategy has been implemented, starting this year to treat Menton's plane trees" (*Ouest-France*, 2012).

10 English translation: "[...] Biological and *ecological control*, a stable goal, these are some of the rules we have imposed on ourselves" (*Le Progrès*, 2013).

focusing on aspects of the referent and the concept that are particularly adapted to the situation"[11] (Mortureux 1993, p. 119).

Replicating the methodological approach to previously identified units allows us to identify other co-referent compounds, such as *protection naturelle*, *régulation naturelle* or *défense écologique*:

> 9) "[…] le directeur général de Koppert France, leader du biocontrôle (*protection naturelle* des cultures) […]"[12] (*Ouest-France*, 2016).
>
> 10) "Pour favoriser les processus de *régulation naturelle*, des bandes enherbées sont installées autour des cultures"[13] (*Ouest-France*, 2017).
>
> 11) "[…] la ville qui a été précurseur dans le domaine de la *défense écologique* des rosiers contre les pucerons"[14] (*Ouest-France*, 2010).

In the first two examples, the adjective *naturelle* is introduced and associated with the predicates *protection* and *régulation*. This qualifier refers to an element pertaining to nature, as opposed to a modified or processed one. In this case, it echoes the second definition of the term *biologique*, which we have previously seen (meaning chemical-free).

The lexical unit *régulation*, which is semantically close to the term *contrôle*, can be explained with a non-redundancy strategy, which is based on the synonymy between terms.

Likewise, in the third example above (11), we observe the semantic proximity between *défense* and *protection*. Thus, the term *défense écologique* implicitly refers to the biological protection of roses. On the basis of synonymous relations within a specific lexical field, the list of designations seems unlimited. For example, we note the emergence of the lexical field of conflict or struggle (*lutte*, *protection*, *défense*, etc.), which reveals another process that is relatively common in popular science, the "conceptual metaphor"; according to the theory of Lakoff and Johnson (1980), this

---

11 Original French version: "Le terme (spécialisé) n'apparaît plus nécessairement – en dépit de sa précision, de son appropriation – comme le plus juste en toutes circonstances, car certains de ses reformulants peuvent avoir l'avantage de focaliser des aspects du référent et du concept particulièrement adaptés à la situation d'énonciation (Mortureux 1993, p. 119).

12 English translation: "[…] the managing director of Koppert France, leader in biocontrol (*natural protection* of crops) […]" (*Ouest-France*, 2016).

13 English translation: "To promote *natural regulation* processes, grass patches are installed around crops" (*Ouest-France*, 2017).

14 English translation: "[…] the city that has been a forerunner in the field of *ecological defense* of roses against aphids" (*Ouest-France*, 2010).

is the use of terms and the conceptual structure of a source domain to describe another – the target domain. This process can be illustrated here by the "warlike" connotation of the noun *lutte*, referring to a "violent opposition between two adversaries", which is likely to be at the origin of the conceptual metaphor of conflict, as illustrated in the following example:

> 12) "Une *guerre biologique* dont l'arme fatale s'appelle *Torymus sinensis*"[15] (*La Montagne*, 2016).

Thus, by virtue of discursive analogies, press articles are enriched with lexical forms of various rhetorical interests (notably non-redundancy and simplification). This may also relate to what Dury and Lervad call the "semantic elasticity" of these terms, in a situation of "circumstantial synonymy"[16] (Dury and Lervad 2008: 71). A "*guerre biologique*" thus refers to a struggle between living organisms to protect plants. Despite the discursive strength of simplifying the explanation of the processes involved, it nevertheless creates a terminological and cognitive ambiguity, which will be explored in greater depth in Chapter 7.

The lexical combinations observed in media discourse can even extend beyond two elements. Terms composed of three lexical units are also identified in corpus, as in the two examples below:

> 13) "Esquive WP est la première solution de *lutte biologique naturelle* homologuée par l'AFSSA"[17] (*Ouest-France*, 2010).

> 14) "Nous avons surtout réintégré des insectes locaux au cœur du parc [...]. C'est une *protection biologique naturelle*"[18] (*Ouest-France*, 2010).

In the previous example, the two adjectives, *biologique* and *naturelle*, are associated; their interpretation is left up to the reader.

Lastly, other co-referential designations that have been identified in the French corpus focus on specific aspects of plant protection methods:

---

15 English translation: "A *biological war* whose lethal weapon is called *Torymus sinensis*" (*La Montagne*, 2016).

16 Original French version: "*L'élasticité sémantique de ces termes, en situation de 'synonymie de circonstance'*" (Dury and Lervad 2008, p. 71).

17 English translation: "Esquive WP is the first *natural biological control* solution approved by AFSSA" (*Ouest-France*, 2010).

18 English translation: "Most importantly, we've put local insects back into the heart of the park [...]. It is a *natural biological protection*" (*Ouest-France*, 2010).

15) "Dans ce type de *lutte parasitaire*, le principe veut que plus il y a de ravageurs, plus les auxiliaires sont actifs"[19] (*La Nouvelle République du Centre*, 2013).

16) "Ces options de *lutte antiparasitaire* incluent [...] l'application d'agents de lutte biologique"[20] (*L'Humanité*, 2017).

17) "[...] et assurer la *lutte sanitaire* sont des pensées quotidiennes pour ces chefs d'orchestre à la main verte"[21] (*Ouest-France*, 2011).

Extracts 15 and 16 contain two nominal groups that both refer to biological control, whilst being built on semantically opposed adjectives (*parasitaire* and *antiparasitaire*). The two terms describe the target of the control process (parasites), and not its *modus operandi* (by living organisms). Extract 17 introduces another co-referential phrase: "*lutte sanitaire*" ("sanitary control"); this describes the purpose of the control (for plant health). These lexical choices may arise from cognitive variation, and involve the notion of perspective or the multidimensionality mentioned earlier. They represent what Mortureux (1994, p. 222) calls "non-synonymous co-referential units"[22], based on terms with no inherent lexical relationship.

Before presenting the synthesis of the French designation paradigm, the designations that have been identified in the English corpus will now be introduced.

5.1.2. *Control methods in English*

This section presents the designation paradigm in relation to the unit *biocontrol* (or *biological control*) in English. The following corpus extract is an illustration of the relatively classical use of this term:

18) "The only realistic, affordable answer is *biological control* – the release of a natural enemy" (*The Daily Telegraph*, 2011).

Textual data analysis allows us to identify three other predicates that can be substituted for the main term of the phrase: *treatment*, *method* or *solution*. The following is an example:

---

19 English translation: "In this type of *parisitic control*, the more pests there are, the more active control agents there are" (*La Nouvelle République du Centre*, 2013).

20 English translation: "These *anti-parasitic control* options include [...] the application of biological control agents" (*L'Humanité*, 2017).

21 English translation: "[...] and ensuring *sanitary control* are daily thoughts for these green-thumbed leaders" (*Ouest-France*, 2011).

22 From the French "*unités coréférentielles non synonymes*".

19) "The benefits of *biological* or *'natural' methods* [...]" (*The Daily Telegraph*, 2016).

Based on these substitute predicates (such as *method*), and in the same way as for the French language, we find alternative adjectives, such as *natural*. By using the combination paradigm, we can then identify new collocations in discourse (such as *natural control* or *organic control*):

20) "[Bug] collection will [...] be a resource for scientists who study *natural controls* on the environment" (*The Guardian*, 2017).

21) "Leatherjacket Killer, which is an *organic control* containing eelworms" (*The Sun*, 2016).

The term *organic* has two meanings (according to both American and British reference dictionaries)[23]: relating to living beings (plants or animal) and/or relating to products grown without fertilizers or pesticides. This term therefore presents the same duality as the French term *biologique* (or the English term *biological)*, and is a synonym of the term *natural* when used with the second meaning, which refers to products being grown without chemical products.

A set of more complex noun phrases also complete the designation paradigm in English, based on three lexical units:

22) "*Natural biological control* is all part of the garden's ecosystem" (*The Times*, 2017).

23) "Spiders are particularly crucial in organic farming, which relies heavily on *biological pest control*" (*Washington Post*, 2014).

From a discursive point of view, the different terms above refer to the same processes or methods. However, from a semantic and cognitive point of view, the terms (*biological*, *natural* or even *organic*) sub-categorize the methods according to their intrinsic characteristics, their type or mode of action, without the reasons for the denominative variation always being obvious (or explicit).

As in French, other terms are used to form co-referential compounds, based on similar semantic features, such as *sustainable*, *environmentally friendly* or *safe*:

24) "The ante has been upped in the past year or two about the need for *environmentally friendly pest control*" (*The Daily Telegraph*, 2016).

23 See: https://www.wordreference.com/definition/organic.

These adjectives also sub-categorize the control method, but from the perspective of its environmental impact.

As in French, the discourse is enriched by analogies, using a very common lexicon. The conceptual metaphor of conflict (struggle) is also very present in this corpus:

> 25) "[Professor X] sought to refine *biological warfare methods* of pest eradication to ensure that they caused as little damage as possible to the environment" (*The Daily Telegraph*, 2011).

> 26) "Brian Spencer, president of family-owned Applied Bio-nomics, says that support from the public sector is crucial to research and the growth of the *biological weapons sector*" (*Financial Times*, 2019).

Other elements on the syntagmatic axis bring a new point of view, introducing either the agent used for biological control (hereafter, parasites, in extract 27) or the purpose of this control (aphids, in 28):

> 27) "The best option for whitefly is to introduce Encarsia, a tiny *parasitic biological control*" (*The Guardian*, 2015).

> 28) "But it was still introduced into parts of Europe in another attempt at *aphid control*" (*The Independent*, 2018).

Whether it is a question of discursive, functional, cross-linguistic or cognitive variation, the language used by journalists is enriched by synonymic uses that are "undoubtedly superfluous, but which are making their way"[24] (Tolédano and Candel 2006, p. 181). A summary of extracts and contexts is given below – in both languages and corpora.

5.1.3. *Summary of method designations*

Table 5.1 presents all the designations that have been identified in the corpus to refer to biological control methods. This is not a translation from one language to another, but rather denominative variations that have been attested in each language, based on the same methodology.

24 Original French version: "*Sans doute superfétatoires, mais qui font leur chemin*" (Tolédano and Candel 2006, p. 181).

	French language	English language
Generic designation (origin/composition)	***Lutte*** *Lutte biologique* *Lutte naturelle* *Lutte biologique naturelle* *Lutte bio* *Lutte microbiologique* ***Protection*** *Protection biologique* *Protection biologique naturelle* *Protection biologique contrôlée* *Bioprotection* ***Contrôle*** *Biocontrôle* *Contrôle biologique* ***Régulation/solution/technologie*** *Régulation naturelle* *Solution biologique* *Technologie biologique*	***Control*** Biological control Natural control Organic control Natural biological control Biocontrol ***Method*** Biological method Natural method ***Treatment*** Biological treatment ***Solution*** Biological solution
Impact on biodiversity	***Lutte*** *Lutte écologique* *Lutte verte* *Lutte raisonnée* ***Défense*** *Défense écologique* ***Contrôle*** *Contrôle écologique*	***Control*** Ecologically-friendly pest control Environmentally-safe pest control Sustainable (natural) control Safe biological control ***Method*** Environmentally-sound methods of control ***Solution*** Environmentally-friendly solution
Objective/Specificity	***Lutte*** *Lutte sanitaire* *Lutte parasitaire* *Lutte antiparasitaire* *Lutte directe* *Lutte individuelle*	***Control*** Parasitic biological control Aphid (biological) control Precision control Specific control
Analogies	*Guerre biologique* *Parade biologique*	Biological warfare Biological weapon sector

Table 5.1. *Designation paradigm for control methods (French/English)*

The first column on the left specifies the sub-categorization suggested for each term, following the semantic analysis.

The next section focuses on the names and designations for the variety of biocontrol agents and products.

5.2. Designations of agents and plant protection products

In this section, we present the denominative variations related to biocontrol agents and products, in the two corpora.

5.2.1. *Biocontrol agents (insects)*

This section concerns biocontrol agents, notably insects, which are the most frequent control agents mentioned in media discourse.

5.2.1.1. *Designations in French*

Biological control agents are generally referred to as "*auxiliaires*" in French[25], and act against pests to protect plants:

> 29) "Ces dernières années, plusieurs municipalités ont annoncé avoir recours à ces '*auxiliaires de lutte biologique*'"[26] (*La Croix*, 2010).

The term *parasite*, from specialized discourse, also refers to control agents:

> 30) "Grâce à un projet de lutte biologique "classique" (introduction d'un *parasite* de *D. citri*), le "dragon jaune" a été totalement éradiqué sur l'île"[27] (*Le Figaro*, 2017).

The parasite can also be referred to as a *prédateur (predator)* that acts against plant pests. Neither of these two terms has a negative connotation, since the control agent is considered as a protector and useful for biological control:

---

25 The definitions of the terms in French can be viewed on the Vegephyl website (see: https://www.vegephyl.fr/publications/bases-de-donnees/base-repertoire-terminologie), as well as in the glossary at the end of the book.

26 English translation: "In recent years, several municipalities have announced the use of these *biological control agents*" (*La Croix*, 2010).

27 English translation: "Thanks to a 'classical' biological control project (introduction of a *D. citri parasite*), the 'yellow dragon' has been completely eradicated on the island" (*Le Figaro*, 2017).

31) "Aujourd'hui, la reproduction et l'élevage du *prédateur* de l'insecte ravageur permettent d'entreprendre avec succès des opérations de lutte biologique"[28] (*Ouest-France*, 2015).

Media discourse also uses more specialized terms to designate control agents (*parasitoides, antagonistes, hyperparasites*). In this case, their interpretation by a non-expert reader is generally achieved through lexico-syntactic or metalinguistic markers (parenthesis, comma, conjunction, etc.):

32) "Grâce à 10 à 12 lâchers annuels d'auxiliaires (insectes *prédateurs* ou insectes *parasitoïdes*) [...]"[29] (*La Nouvelle République du Centre*, 2017).

33) "L'utilisation de *prédateurs, antagonistes ou hyperparasites* pour lutter contre une autre espèce ravageuse [...]"[30] (*Le Monde*, 2014).

The following are designations of biocontrol agents in English.

5.2.1.2. *Designations in English*

Insects can be described as biological control agents:

34) "The insects are considered *biological control agents* by invasive plant experts" (*The New York Times*, 2013).

The terms *parasite* and *predator* are also used in English:

35) "The history of biocontrol – where *predators, parasites* or *pathogens* are released into a habitat to bring an invasive under control" (*The Financial Times*, 2016).

36) "Treat moderate infestations with the *biological predator* Phytoseiulus persimilis" (*The Daily Telegraph*, 2013).

Other lexical units fit in the designation paradigm, such as *controller*:

37) "A jeweled *biological controller* [...] dock beetles are the organic farmer's friend" (*The Guardian*, 2015).

---

28 English translation: "Today, the reproduction and breeding of the *predator* of insect pests allow for successful biological control operations" (*Ouest-France*, 2015).

29 English translation: "Through 10 to 12 annual releases of auxiliaries (*predatory* insects or *parasitoid* insects) [...]" (*La Nouvelle République du Centre*, 2017).

30 English translation: "The use of *predators, antagonists* or *hyperparasites* to control another pest species [...]" (*Le Monde*, 2014).

Note the introduction of the specialized term *parasitoid*, which refers to the control agent:

> 38) "Van Driesche is researching *parasitoids* – organisms akin to *parasites* but with the distinction of killing their host" (*The Financial Times*, 2016).

The summary of the designations identified in the corpus are presented below.

5.2.1.3. *Summary of control agent designations*

	French language	English language
Generic designation	***Agent*** *Agent de lutte* *Agent de lutte biologique* *Agent de biocontrôle* ***Auxiliaire*** *Auxiliaire de lutte Biologique* *Auxiliaire de cultures* *Auxiliaire de jardinage* ***Prédateur*** *Insecte prédateur* ***Parasite*** *Insecte parasite* *Hyperparasite* ***Parasitoïde*** *Insecte parasitoïde* *Auxiliaire parasitoïde* ***Organisme*** *Macro-organisme* *Organisme naturel*	***Agent*** Biocontrol agent Biological control agent Control agents Insect biocontrol agents ***Organism*** Microscopic organisms Soil organisms Parasitoid organism Control organism ***Predator*** Biological predator Micro-predators Natural soil predator ***Parasite*** Natural parasites ***Microbe*** Disease-fighting microbe ***Parasitoid*(s)** ***Control*(s)** Biological controls Natural controls Controller
Role/Impact	*Insecte protecteur* *Insecte bénéfique* *Insecte utile* *Insecte nécessaire* *Antagoniste* *Ennemi (naturel)*	Auxiliary insect Insect enemy
Examples	*Mini-guêpes* *Micro-guêpes* *Trichogramme* *Encarsia (guêpes parasitaires)* *Punaises prédatrices*	***Wasp*** Micro wasp Parasitic wasp Encarsia Formosa

Table 5.2. *Designation paradigm for biological control agents*

Table 5.2 presents all of the designations of pest control agents (macro-organisms) in context and illustrates the specificities of each language. The column on the left shows their sub-categorization.

We now turn to biocontrol products in both languages, and corpora.

5.2.2. *Biocontrol products*

Media discourse either refers to products using generic terms (*produit, products, solutions*) or more specific terms (*bioinsecticide, biopesticide*, etc.). The French designations are given below.

5.2.2.1. *Designations in French*

The following extracts illustrate the different products or solutions involved:

> 39) "On peut intervenir avec du *produit de bio contrôle*, du Bacillus thuringiensis"[31] (*Ouest-France*, 2017).

> 40) "NPP […] produit des *solutions de biocontrôle* à base d'huiles essentielles"[32] (*Les Echos*, 2014).

Other co-referential terms in French are built on the suffix *cide*, indicating the purpose of the control process, which, in this case, is killing pests:

> 41) "La culture bio, encore peu développée, remplace les pesticides par des *insecticides naturels* (feuilles de neem)"[33] (*Le Progrès*, 2010).

> 42) "La lutte microbiologique à l'aide de *bio insecticide* comme le bacille de Thuringe"[34] (*Ouest-France*, 2016).

> 43) "[…] Pour utiliser une souche de champignon capable de sécréter un *biocide* très efficace contre ses congénères oomycètes pathogènes"[35] (*Le Figaro*, 2018).

31 English translation: "We can intervene with a *biocontrol product*, Bacillus thuringiensis" (*Ouest-France*, 2017).

32 English translation: "NPP […] produces *biocontrol solutions* based on essential oils" (*Les Échos*, 2014).

33 English translation: "Organic farming, which is still underdeveloped, replaces pesticides with *natural insecticides* (neem leaves)" (*Le Progrès*, 2010).

34 English translation: "Microbiological control using *bioinsecticide* such as Thuringian bacillus" (*Ouest-France*, 2016).

Press discourse regularly makes use of examples to illustrate control methods and products. These are then introduced by lexico-syntactic markers (which is shown here in the form of a colon or a parenthesis):

> 44) "Autre méthode de lutte biologique: la pulvérisation de *biotoxines*, produites par la bactérie *Bacillus thuringiensis*"[36] (*La Croix*, 2019).

> 45) "Utilisation [...] de décoction à base de substances naturelles (*purin d'orties*), ou de médiateurs chimiques (*piège à phéromones, confusion sexuelle*) [...]"[37] (*La Nouvelle République du Centre*, 2017).

As illustrated by the second extract, some examples are presented in the form of a list. This list is generally based on a generic term[38] from official regulatory texts: biotoxins are introduced by the generic term "biological control method" in the first example (44); in the second example (45), "nettle compost" is associated with natural substances, pheromone traps, and sexual confusion to chemical mediators.

The following focuses on the designation paradigm in English.

5.2.2.2. *Designations in English*

The term *biopesticide* (or its expanded form *biological pesticides*) is commonly used to refer to biocontrol products:

> 46) "Scientists are working on *biopesticides*, such as viruses that naturally attack the armyworms" (*The Guardian*, 2017).

> 47) "[...] Making *biological pesticides* – mainly, live bacteria that can kill the worms and caterpillars" (*The Washington Post*, 2017).

As Eilenberg et al. already pointed out, the term *biopesticide* is based on the chemical model: "Insect pathologists tend to pick up words from the pesticide industry and speak of 'biopesticides' (for example, the use of *Bacillus thuringiensis*)" (Eilenberg et al. 2001, p. 388).

---

35 English translation: "[...] To use a strain of fungus that is capable of secreting a *biocide* that is highly effective against its pathogenic oomycete counterparts" (*Le Figaro*, 2018).

36 English translation: "Another method of biological control: spraying biotoxins, produced by the bacterium *Bacillus thuringiensis*" (*La Croix*, 2019).

37 English translation: "The use [...] of decoctions based on natural substances (*nettle purin*), or chemical mediators (*pheromone traps, sexual confusion*) [...]" (*La Nouvelle République du Centre*, 2017).

38 "*Amorce*" in French (Rebeyrolle 2014).

However, to ensure that this term is clearly different from chemical pesticides, the journalist proposes several reformulations. In the following example, the reader understands that "natural pesticide" is correlated to biopesticide, based on the two successive sentences:

> 48) "Scientists who fed bees with high doses of the *bio-pesticide* Hv1a/GNA observed only a very slight impact on survival [...]. The *natural pesticide* could be a safer alternative" (*The Times*, 2014).

The term *all-natural* is also found in corpus to create another correlated term, in which the lexical unit *all* acts as an intensifier for the term as a whole:

> 49) "The Aim-listed firm has devised an *all-natural pesticide* that kills botrytis [...], derived entirely from plants" (*The Washington Post*, 2014).

From a semantic standpoint, the two terms (*biological* and *natural*) sub-categorize pesticides according to their intrinsic characteristics (by living organisms or by methods existing in the natural world).

As in French, other terms (such as *safe* or *environmentally sustainable*) also sub-categorize products according to their assumed environmental impact:

> 50) "Plague of caterpillars threatening food crisis may be halted with *safe pesticides*" (*The Guardian*, 2018).

> 51) "Government also must minimize regulatory bottlenecks to promote the development of more *environmentally sustainable insecticides*" (*The Wall Street Journal*, 2017).

In the paradigm of semantically equivalent modifiers, we also find the terms *environmentally safe* and *environmentally friendly*. Other qualifiers are to be interpreted in terms of their impact on biodiversity, insects in particular, as shown in this following example:

> 52) "By licensing Eden's fully regulated, *bee-friendly pesticides* [...]" (*The Daily Telegraph*, 2014).

The adjective *bee-friendly* has the same discursive function as other similar qualifiers identified in corpus, such as *bee-safe*, *insect-friendly* and *wildlife-friendly*.

None of these excerpts provide a definition for biopesticides, but the reader can, from textual cues or co-textual elements, infer that these pesticides are "produced" from natural elements, and are therefore generally harmless to biodiversity (the environment and animals).

Below is a summary of the designations identified in the double corpus.

5.2.3. *Summary of product designations*

Table 5.3 shows the variety of terms that refer to the various (biological) control products.

	French language	English language
Generic designation (origin/composition)	*Produit de biocontrôle* *Solution de biocontrôle* *Substance de biocontrôle* *Produit biologique* *Pesticide d'origine naturelle* *Insecticide* *Insecticide biologique* *Insecticide naturel* *Bio insecticide* *Biopesticide (bio pesticide, bio-pesticide)* *Fongicide, biofongicide* *Produit larvicide* *Produit phytopharmaceutique de biocontrôle* *Produit organique* *Biotoxine* *Biocide*	Biological pesticide Biopesticide Bio-pesticide Natural pesticide Natural product All-natural pesticide Fungi Bacteria Phytoseiulus persimilis (predatory mite) Bacillus thuringiensis (Bt)
Impact on biodiversity	*Produit inoffensif* *Solution naturelle et biodégradable* *Solution durable* *Produit sans danger*	Safe pesticide Environmentally-friendly pesticide Environmentally-sustainable alternatives/insecticides Insect-friendly pesticide Wildlife-friendly pesticide Bee-friendly/bee-safe pesticide Sustainable insecticide Environmentally-sustainable
Objective/Specificity	*Produit sélectif*	Specific pesticide Selective pesticide

Table 5.3. *Designation paradigm for biocontrol products*

This proliferation of terms is even more significant when considering other combinations in which a modifier becomes the main term of a new compound; this is the case for the modifier *alternative* in French and English, which becomes the predicative noun, for example, in *ecological alternative*, *green alternative*, *sustainable alternative* or *safer alternative*, completing the paradigm of potential designations.

As a conclusion to this first part on textual analysis, and following Moirand's example, "naming is not only about representing what one is talking about, it is, in this case, also about designating and characterizing *for others*, i.e., for the different classes of potential recipients, in particular the readers of the mainstream press"[39] (Moirand 2007, p. 31).

In the next chapter, the characterization of these discourse objects will be examined.

39 Original French version: "Nommer, ce n'est pas seulement représenter ce dont on parle, c'est également désigner et caractériser, *pour les autres*, ici pour les différentes classes de destinataires potentiels, y compris les communautés concernées, que sont les publics de la presse ordinaire" (Moirand 2007, p. 31).

6

Characterization of Discourse Objects

Common sense is something
the world has in common[1].

Jean-Claude Anscombre (in Cislaru and Nyckees (2019))

In this chapter, we shall delve further into the description of methods, agents and products that have been previously identified, based on the methodology presented in Chapter 3. Our objective is to study the lexico-discursive structures in which they appear, as well as the semantic relations they share with other elements in the corpus.

In press articles related to this innovative field, there are very few defining contexts to speak of, which involve metalinguistic definition processes using verbal markers (such as *signifier, désigner, s'appeler* in French[2] (Mortureux 1993; Riegel 1987)). We have already demonstrated this for the terms *lutte intégrée* (*integrated pest management*) (Ledouble 2021) and *biocontrôle* (*biocontrol*) (Ledouble and Cadel 2022). However, an "explanation process", specific to press articles, allows the journalist to characterize the elements in discourse. We present the results of our analyses in the double corpus below.

6.1. Crossed perspectives on methods and products

In this section, we gather the results of the analysis of the French and English corpora together, as they present similar discourse characteristics. We intend to

1 Original French version: "*Le sens commun est la chose du monde la mieux partagée*".

2 Equivalent of the verbs *to mean*, *to designate*, *to name* in French.

focus on the different semantic relations that characterize control and biocontrol methods and products presented in Chapter 5.

6.1.1. *Complementarity of methods and products*

An analysis of textual data shows that biological control methods can be presented as complementary to chemical control methods (based on the perspective of Integrated Pest Management as recommended in the legal European Directive). Some corpus extracts convey this complementarity principle through "coordinated antonymy", and lexico-syntactic markers indicate this type of discursive relations:

> 53) "Tous nos produits, *qu'ils soient de* synthèse, d'origine naturelle utilisés en bio (cuivre, soufre, etc.) *ou encore de* bio-contrôle comme les phéromones, font l'objet d'une batterie de tests"[3] (*L'Humanité*, 2015).

> 54) "Control is achieved […] *either with* a biological control based on nematodes […], *or* chemical control" (*The Daily Telegraph*, 2015).

The linguistic markers *qu'ils soient de/ou encore de* (*whether they be/of* in English), and *either/or* allow the reader to understand that, in both cases, we are dealing with products or control methods for plant protection which, as such, have common ground. Their differences are then neutralized to the benefit of their similarities: the first context (53) conveys the fact that biocontrol products are subject to a series of tests, in the same way as other plant protection products; the second context (54) expresses the fact that the control process can be carried out through biological or chemical methods.

Other excerpts recall the principle of complementarity between methods, based on lexical markers:

> 55) "Ces techniques [de biocontrôle] s'utilisent rarement seules, plutôt dans le cadre de ce qu'on appelle la 'protection intégrée', c'est-à-dire *en complément de* traitements traditionnels"[4] (*La Croix*, 2016).

> 56) "Before specific vine weevil killer treatments (*both natural and chemical*) were developed […]" (*The Daily Telegraph*, 2016).

3 English translation: "All of our products, whether they be synthetic, of natural origin used in organic farming (copper, sulfur, etc.) or bio-control products such as pheromones, are subject to a battery of tests" (*L'Humanité*, 2015).

4 English translation: "These [biocontrol] techniques are rarely used alone, but rather as a part of 'integrated protection', i.e. *as a complement* to traditional processes" (*La Croix*, 2016).

The markers *en complément de* (*as a complement to* in English) or *both* show that the different processes are related, and constitute two co-hyponymic methods of plant protection. Cruse (1986) had already accounted for this apparent contradiction in the antonymy relation:

> This paradox of simultaneous difference and similarity is partly resolved by the fact that opposites typically differ along only one dimension of meaning: in respect of all other features they are identical, hence their semantic closeness (Cruse 1986, p. 197).

Beyond their complementarity, these methods are also shown to be compared in press discourse, as we will now present.

6.1.1.1. *Comparison between methods*

Complementarity is often associated with a comparison, related, for the most part, to the impact of one of the methods on the environment and biodiversity. This can be expressed by using classical comparison markers, such as *plus* or *moins* (*more* or *less* in English), and by the comparative forms of adjectives in English (such as *wider*):

> 57) "Ce genre de programme mixte [*biocontrôle* + *lutte chimique*] est *plus respectueux* de l'environnement"[5] (*La Croix*, 2016).

> 58) "Reaching for the spray [chemical pesticides] would only contribute to the *wider threat* to bees and other "friendly" insects" (*The Independent*, 2014).

In the first example (57), the comparison is between a chemical-only method and a hybrid method (*biocontrôle* + *chimique*, equivalent to *biocontrol* + *chemical* in English), which is considered to be more environmentally friendly. In the second example (58), the noun group *wider threat* expresses the fact that chemicals pose a greater threat to the environment than biocontrol products.

The effectiveness of control methods is an argument that is also regularly emphasized through comparisons. In the following extracts, the contrast in terms of effectiveness is achieved with the help of the marker *plus* associated with *moins* in French (*more* or *less* in English), and *just as* in English:

5 English translation: "This kind of mixed program [*biocontrol* + *chemical control*] is *more* environmentally friendly" (*La Croix*, 2016).

59) "Parmi ces auxiliaires, les coccinelles, très friandes de pucerons et bien *plus efficaces et moins polluantes* que nombre de produits phytosanitaires"[6] (*La Montagne*, 2016).

60) "Eden's botrytis product, 3AEY, is not only derived entirely from plants, it's also cheap to manufacture and *just as effective* as toxic chemical variants" (*The Daily Telegraph*, 2014).

In both languages, it is noted that "natural" methods are occasionally presented as being just as effective as their chemical counterparts.

The financial issue is also highlighted in many contexts, often in relation to environmental protection. However, this comparison does not necessarily lead to the same observations or conclusions. The following two examples show that biological control has a potentially significant cost (in 61), and in the following example (62), alternative methods appear as potentially less expensive solutions to chemical products (*pesticides*):

61) "Si cette manière de lutter contre les parasites est *plus naturelle* et *plus efficace* à long terme, elle a *tout de même un coût*: 4 000 euros par an, sans compter le *temps passé* et le *matériel*"[7] (*La Nouvelle République du Centre*, 2017).

62) "The high cost of pesticides [...] was having an adverse effect on family incomes. They were motivated to find *safer* and *cheaper* methods of cotton production" (*The Guardian*, 2018).

Other characteristics are highlighted in press articles, combining various comparisons (economic, ecological, etc.):

63) "On doit créer une agriculture qui soit *intensive en emploi plutôt qu'en engrais* et produits phytosanitaires"[8] (*Ouest-France*, 2017).

---

6 English translation: "Among these auxiliaries are ladybugs, which are very fond of aphids, and are *much more effective and less polluting* than many plant protection products" (*La Montagne*, 2016).

7 English translation: "Although this way of controlling pests is *more natural* and *more effective* in the long-term, *it is still costly*: 4,000 euros per year, not counting the *time spent* and the *equipment*"[7] (*La Nouvelle République du Centre*, 2017).

8 English translation: "We must create an agriculture that is job intensive, *rather than* fertilizer and plant protection product intensive" (*Ouest-France*, 2017).

64) "Backers of conventional agriculture say they have *the same goals as* organic farmers, including better environmental stewardship" (*The Wall Street Journal*, 2015).

In these two excerpts, the comparison markers (*plutôt*, *same as*) emphasize different aspects. On the one hand, differences in the objectives (promoting the use of fertilizers or chemical products), and, on the other hand, their convergences (environmental protection).

Some extracts fall under a sub-categorization of comparative antonymy (defined by Jones as "distingued antonymy"), which emphasizes a contrast between two opposing elements in discourse. In the following extracts, it is the marker *entre* in French and *balance* in English that conveys this contrast:

65) "C'est là tout l'enjeu de l'activité de notre secteur: permettre aux agriculteurs de ne pas avoir à choisir *entre* la qualité et la sécurité de leurs récoltes, *ou* la préservation de la faune et la flore"[9] (*L'Humanité*, 2015).

66) "Our purpose is to *balance* serving the public health *and* environment with enabling farmers' access to innovation" (*The New York Times*, 2017).

The comparison draws attention to specific distinctions between the different methods. In example 65, the quality and safety of production are guaranteed by chemical treatment, as opposed to a more natural treatment. In example 66, the distinction is made on health, the environment and innovation.

Other extracts move away from the complementarity between these methods or products, to emphasize the transition from one to another.

6.1.1.2. *Transition between methods*

A great number of articles express the need for a transition between these methods. This is conveyed through different types of lexical markers, either from the lexical field relating to the transition imposed by legal texts (*transition*, *passer du… au*, etc.), as in the French example below (67), or associated with alternative agricultural conceptions (or practices) (*rethink*, *outdated*, etc.), as in the English example (68):

---

9 English translation: "That is what our industry is all about: not having to make farmers choose *between* the quality and safety of their crops, or the preservation of wildlife" (*L'Humanité*, 2015).

67) "Cette *transition* vers l'agroécologie, qui va permettre une diminution des pesticides, est désormais engagée"[10] (*La Tribune*, 2016).

68) "It is time for a major initiative to *rethink* agriculture globally" (*The Independent*, 2018).

Transition between methods is also expressed by the predicates *diminuer* (*reduce* in English) and *remplacer* (*replace* in English), associated with pesticides, as a distinctive feature of this type of antonymy:

69) "Les substances de synthèse ont *beaucoup diminué, remplacées par* des macro-organismes ou des bactéries"[11] (*Les Échos*, 2017).

70) "If we increasingly *replace* pesticides with biological pest control then we can grow food more sustainably" (*The Financial Times*, 2013).

The goal is to (progressively) reduce chemical plant protection products, without of course having the possibility or the intention to eliminate them immediately, as some articles rightly recall:

71) "Deux viticulteurs qui ne sont pas bio, mais qui sont néanmoins convaincus de l'utilité de *diminuer, autant que faire se peut*, l'usage des produits phytosanitaires"[12] (*Le Figaro*, 2011).

72) "[...] Producers [...] are starting to *reduce* their reliance on chemicals and *develop alternative controls* for fruit flies" (*The Guardian*, 2016).

So far, these extracts have highlighted the progressive reduction of plant protection products in favor of other methods, as suggested by the 2009 European directive (see section 4.1). Other articles put more emphasis on a more radical opposition between plant protection methods and products. These are presented below.

10 English translation: "This *transition* to agroecology, which will allow for a decrease in pesticides, is now underway" (*La Tribune*, 2016).

11 English translation: "Synthetic substances have *greatly been reduced*, and have been *replaced* by macro-organisms or bacteria" (*Les Echos*, 2017).

12 English translation: "Two wine producers who are not organic, but who nevertheless see the usefulness of *reducing, as much as possible*, the use of plant protection products" (*Le Figaro*, 2011).

6.1.2. *Opposition between methods and products*

In our double corpus, we also find a variety of semantic structures pinning natural methods/products against chemical methods/products, according to three different perspectives: their status (biological vs. chemical), their harmfulness (healthy vs. dangerous), and a socio-temporal consideration related to their use (innovative vs. classic). We will now present them in this order.

6.1.2.1. *Biological versus chemical*

First of all, we find negated antonymy, based on predefined operators (*au lieu de, in place of*, etc.) in both corpora:

> 73) "*Au lieu de* pulvériser des produits phytosanitaires, nous utilisons l'apport d'auxiliaires, type coccinelles"[13] (*Le Progrès*, 2014).

> 74) "It [Bti] is now widely used *in place of* chemical pesticides" (*The New York Times*, 2014).

In these extracts, biological control methods and products are generally opposed to "chemical" ones (here referred to as *produits phytosanitaires in French*, and *chemical pesticides* in English). This recurrent antagonism is conveyed through a gradual form of opposition. The following extract expresses a sharp contrast between control methods:

> 75) "À noter que depuis 2008, *plus un seul* produit chimique n'est utilisé dans les serres municipales, qui ont adopté la 'protection biologique intégrée'"[14] (*La Nouvelle République du Centre*, 2017).

While it is possible to highlight the absence of chemical treatment in this specific case of greenhouses, as illustrated in example 75, this cannot necessarily be achieved in a field (due to questions of scale or production conditions). However, this distinction is not always explicit, as press discourse generally emphasizes a radical exclusion of chemical treatments:

> 76) "Par exemple, pour décimer les pucerons, *point besoin* de produits toxiques: ce sont les larves de la mouche noire qui s'en chargent"[15] (*La Voix du Nord*, 2014).

---

13 English translation: "*Instead* of spraying plant protection products, we use auxiliaries, such as ladybugs" (*Le Progrès*, 2014).

14 English translation: "Note that since 2008, not a single chemical *has been* used in municipal greenhouses, which have adopted 'integrated biological protection'" (*La Nouvelle République du Centre*, 2017).

77) "Extracts from need seed – a bio-pesticide – will be prepared and sold. *Instead of* the quick fix of chemicals, these farmers [...]" (*The Guardian*, 2018).

The information is regularly transmitted through reported speech, as in examples 78 and 79. The journalist then endorses the message, and quotation marks are not systematically present:

78) "*Exit* le tout chimique. *Place aux* alternatives aux pesticides. C'est le message des acteurs du biocontrôle"[16] (*Libération*, 2014).

79) "Conventional pest control means such as crop dustings of pesticides were *out of the question*" (*The Washington Post*, 2012).

The lexico-syntactic markers (*exit le, out of the question*) emphasize the radical end of a (chemical) system that *gives way* (*place à* in French) to another, more natural treatment, following a dichotomous and exclusive opposition.

The ban on these products is regularly recalled:

80) "Les produits phytosanitaires ont été *totalement bannis* et remplacés par de la protection biologique intégrée"[17] (*Ouest-France*, 2012).

81) "Lawmakers in Maryland even tabled a bill to *ban* chlorpyrifos, a pesticide deemed unsafe by the Environmental Protection Agency" (*The New York Times*, 2018).

The lexical markers express the fact of limiting, reducing and decreasing chemical plant protection products as well as banning them, in both corpora. We will reconsider the issues linked to this opposition in Chapter 7. Let us present another contrast that is prevalent in the press.

6.1.2.2. *Safe versus dangerous*

Biological methods are regularly highlighted for their closeness to nature, in opposition to the harmfulness (even toxicity) of other methods or products:

15 English translation: "For example, to eradicate aphids, *there is no need* for toxic products: black fly larvae will take care of it" (*La Voix du Nord*, 2014).

16 English translation: "*No more* chemicals. *Make way* for alternatives to pesticides. This is the message of the biocontrol actors" (*Libération*, 2014).

17 English translation: "Plant protection products have been *totally banned* and replaced by integrated biological protection" (*Ouest-France*, 2012).

82) "Le *naturel* revient enfin au galop. Cette fois, c'est décidé: *fini les produits toxiques* qu'on déverse sans compter"[18] (*La Montagne*, 2010).

83) "Supermarkets can say that their flowers use *nature's own defence instead of nasty chemicals*" (*The Daily Telegraph*, 2014).

A lexical field related to a dystopian situation for the environment is recurrent in the corpus (*polluant, contaminé, dangereux, poison, hazardous, toxic*, etc.):

84) "Si l'on exclut les *traitements chimiques polluants* et souvent *plus dangereux* que la chenille, il reste l'échenillage et la lutte biologique"[19] (*La Montagne*, 2010).

85) "They are *organic* – so no *pesticide or poison* is involved" (*The Sun*, 2011).

This lexical field can also be seen in other studies on pesticides in the press (Botero 2021). In this case, media framing quickly shifts from simple doubt about the assumed effect of pesticides, to blaming an "agrotoxic" and biodiversity-destroying product (Botero 2019, p. 112).

Other corpus extracts describe these products with illustrative and connotated designations (here, *granulés bleus*[20] and *lurid pellets*):

86) "Autre indicateur d'un jardin en bonne santé? Le retour des vers luisants qui aiment dévorer les limaces. Avec eux, fini les *granulés bleus*?!"[21] (*Ouest-France*, 2017).

87) "After a lifetime of quick-fix chemicals, are we not rather wedded to the familiar, shamefully careless ritual of spraying unguents and casting about *lurid pellets*"? (*The Daily Telegraph*, 2016).

As Jackiewicz and Pengam (2020) have analyzed (while studying other domains), these expressions or phrases have the discursive aim of "condensing a

18 English translation: "The *natural* way is finally making a comeback. This time, it's been decided: *no more toxic products* will be poured out without careful consideration" (*La Montagne*, 2010).

19 English translation: "If we exclude *chemical treatments* that are *polluting* and often *more dangerous* than the caterpillar, there is still weeding and biological control" (*La Montagne*, 2010).

20 English translation: *Blue pellets*.

21 English translation: "Another indicator of a healthy garden? The return of glow worms that like to devour slugs. With them, no more *lurid pellets*?!" (*Ouest-France*, 2017).

large amount of information about the reference, from its most objective aspects (aim, modus operandi, etc.) to the most subjective aspects (images, values, intensity, etc.), all of it being socially shared"[22]. Thus, in the absence of contextual definitions, the nonexpert reader can infer that chemical products and methods are regularly presented in an antonymic relation with biological pesticides, and portrayed as having a negative impact on biodiversity. However, this view does not always accurately convey the different processes involved, as we will attempt to explain later.

We observe a third form of opposition in the corpus, based on time, and the difference between older and newer plant protection methods and products.

6.1.2.3. *Innovative versus classic*

This form of opposition is characterized by the disparity between current and innovative methods, and more "classic" chemical methods:

> 88) "Si les essais montrent une efficacité de 85 % à 90 % par rapport à des *pesticides classiques*, le coût est bien supérieur"[23] (*Les Échos*, 2014).
>
> 89) "Ce texte donne la possibilité [...] d'utiliser des *produits phytopharmaceutiques classiques* lorsque la survie d'une espèce végétale d'intérêt patrimonial ou biologique est en jeu"[24] (*Le Figaro*, 2017).

According to Le Petit Robert, the French term *classique* refers to that which "conforms to customs, and does not deviate from established rules, from measure"[25]. It is given as a synonym for *traditionnel*, *courant*, *ordinaire* or *standard* (*traditional*, *current*, *ordinary* and *standard* in English). It can be hypothesized that the industrialization of agriculture after the 1930s and the generalized use of synthetic plant protection products, have generated this connotation (the *chemical* era, which has become "*classic*").

---

22 Original French version: "[...] *condenser une grande quantité d'informations sur le référent, de ses aspects les plus objectifs (cible, mode opératoire, etc.) aux plus subjectifs (images, valeurs, intensité, etc.), mais partagées socialement*".

23 English translation: "While trials show 85% to 90% effectiveness compared to *classic pesticides*, the cost is much higher" (*Les Echos*, 2014).

24 English translation: "This text makes it possible [...] to use *classic plant protection products* when the survival of a plant species of heritage or biological interest is at stake" (*Le Figaro*, 2017).

25 Original French version: "[...] *est conforme aux usages, ne s'écarte pas des règles établies, de la mesure*".

Through lexical reformulation (using synonyms or other units belonging to the same lexical field), we can find instances in the corpus where the term *traditionnel* (*traditional*) is equivalent to the term *classique* (*classic*), and therefore to *chimique* (*chemical*):

> 90) "[...] Une agriculture éco-responsable où le "traitement par pesticides est beaucoup plus limité que dans *l'agriculture traditionnelle*"[26] (*Ouest-France*, 2017).

> 91) "Sean Smith, who joined the firm from the *traditional chemical industry* two months ago" (*The Daily Telegraph*, 2014).

Other modifiers demonstrate this same relation to chemical methods, through the use of lexical synonymy, for example, the adjective *conventionnel* (*conventional* in English):

> 92) "Venez échanger sur les pratiques agricoles biologiques, les débouchés en circuit court et les alternatives à *l'agriculture conventionnelle*"[27] (*La Nouvelle République du Centre*, 2014).

> 93) "*Conventional pest control* means such as crop dustings of pesticides were out of the question" (*The Washington Post*, 2012).

We also identify a paradigm of modifiers that are not semantically equivalent to the term *chemical*, but are, however, associated in discourse with this type of method (e.g., *agricultural*), which create discourse compounds that are coreferential to chemical methods:

> 94) "The causes are open to speculation, with *agricultural pesticides* high on the list" (*The Washington Post*, 2018).

Semantically, the term *agricultural* refers to the field of application of pesticides (as opposed to indoor pesticides, such as biocides). However, the reader will understand from the context that the term *agricultural* is contextually interchangeable with *traditional* or *conventional*, the three modifiers referring to (chemical) methods used in agriculture. None of these modifiers can be considered as a synonym.

---

26 English translation: "[...] An eco-responsible agriculture where 'treatment with pesticides is much more limited than in *traditional agriculture*'" (*Ouest-France*, 2017).

27 English translation: "Come and discuss biological farming practices, short circuit outlets and alternatives to *conventional agriculture*" (*La Nouvelle République du Centre*, 2014).

This lexical paradigm referring to what is common or traditional is therefore opposed to what is innovative or future-oriented, and which, in our double corpus, generally refers to biological methods:

> 95) "Éradiquer le parasite à grands coups de pesticides? *Ça, c'était avant, au XXᵉ siècle…* Dès l'an 2000, Villeurbanne a introduit la "lutte biologique"[28] (*Le Progrès*, 2016).
>
> 96) "*L'agriculteur de demain* est agro-écologue, entrepreneur"[29] (*Libération*, 2016).
>
> 97) "[…] A complex political and social problem, not only because of *modern concerns* about pesticide use" (*The Washington Post*, 2010).
>
> 98) "Surprisingly, given our long-standing devotion to *kill-everything chemical pesticides*, it was in the United States that this *precision biological approach* was first employed" (*The Washington Post*, 2018).

These extracts thus contrast modern, innovative (future), digital, intelligent and precision agriculture with industrial, biodiversity-harming agriculture of the past, which is emblematic of chemical "classicism".

6.1.3. *Synthesis of oppositions and binarism*

We are in the situation that Geeraerts et al. (1994) described as "onomasiological variation", in which "the same object may be identified as a member of different categories. […] A particular pair of trousers might be referred to either as a member of the category trousers/pants, or as a member of the subordinate category jeans/blue jeans". The adjectives *classic*, *conventional* and *traditional* refer to the sub-category of chemical methods used in agriculture. They are compared with new methods that are more current, innovative and natural (but not necessarily less intensive). A similar study on the role of adjectives associated with "waste" (Altmanova et al. 2018) also shows the critical significance of the qualifier in the naming and classification function of a term. Based on the useful distinction between transparency and opacity of terms described by Kleiber (1979)[30], the authors

28 English translation: "Eradicating the pest with pesticides? *That was before, in the 20th century…* Since 2000, Villeurbanne has introduced 'biological control'" (*Le Progrès*, 2016).

29 English translation: "*Tomorrow's farmer* is an agroecologist, an entrepreneur" (*Libération*, 2016).

30 Kleiber, G. (1979). À propos de l'ambiguïté. Transparence/Opacité. *Travaux de linguistique et de littérature*, 17, 1, 233–250.

demonstrate the importance of extra-linguistic factors on their interpretation. These allow the creation of sub-categories (of "waste", in the case mentioned), which shows the tenuous boundary between lexicon and pragmatics, as well as the interpretative complexity of nominal groups in the environmental field.

In our study, the reader also relies on extra-linguistic factors, whether they be spatio-temporal elements (Kleiber 2008) or based on the commonly shared opposition between *biological* and *chemical*, thus placing the methods on one side or the other of the evaluation scale. We are facing a form of binary framing that has a discursive interest in terms of scientific mediation, but also presents a certain number of issues, which we will return to in Chapter 7.

We now turn to other strategies for simplifying information.

6.2. Simplified look at methods and products

Lastly, we present two discursive strategies aimed at simplifying explanations of control methods and products: these are either textual or linguistic structures that convey simplicity, or analogies that draw on a broader context, including real-life situations.

6.2.1. *Simplification through (co)textual structures*

There is a set of structures that characterize popular discourse, whose objective is to help the reader understand the processes described. These lexico-syntactic structures are illustrated in the corpus by the following examples: *il suffit de, c'est simple, traduction, put simply, in a few words, for beginners*, etc.

Here are some excerpts that emphasize the simplicity of the processes and the fact that they do not harm the environment:

> 99) "Nous assurons une culture raisonnée avec utilisation de la lutte biologique", indique Jérôme. *Traduction*: "Nous utilisons des bons insectes qui mangent les mauvais"[31] (*Ouest-France*, 2018).

> 100) "*Il suffit d'*ouvrir un tube à essai dans les champs plutôt que de vaporiser des pesticides qui menacent la santé des agriculteurs"[32] (*Le Monde*, 2014).

31 English translation: "'We ensure a sustainable cultivation with the use of biological control', says Jérôme. *Translation*: 'We use good insects that eat the bad ones'" (*Ouest-France*, 2018).

101) "*Put simply*. Biological pest control involves safely harnessing the destructive power of nature's assorted predators and parasites" (*The Daily Telegraph*, 2016).

102) "Biological warfare *for beginners*" (*The Daily Telegraph*, 2016).

We also note the use of an elementary, or even childish lexicon (*bestioles*, *bêtes à bon Dieu*[33], *munchers*, etc.), to describe the processes:

103) "Contre les *petites bêtes* nuisibles, les jardiniers ont recours à d'autres *petites bêtes*. [...] *Vive les bestioles*[34]" (*L'Est Républicain*, April 2014).

104) "La société Koppert, leader mondial en protection biologique intégrée et fournisseur officiel de *bêtes à bon Dieu*"[35] (*La Voix du Nord*, March 2012).

105) "*Biological munchers* who have eaten the unwanted intake" (*The Financial Times*, 2017).

106) "A vine weevil is a *beetle-like creature with a long snout*" (*The Daily Telegraph*, 2017).

A second simplification strategy is the use of analogies.

6.2.2. *Simplification by analogical reasoning*

Various analogies have been analyzed in the media coverage of agroecology, notably those relating to conflict, gastronomy, medicine and, more generally, comparisons to everyday life situations (see Ledouble 2021; Monachon 2021, forthcoming). Below, we present a few illustrations of these analogies from our double corpus, which simplify the explanation of the processes under study.

The most common analogy, after the conceptual metaphor of conflict, is that of food and gastronomy (*croquer les pucerons*[36], just like pepper):

32 English translation: "*Just* open a test tube in the fields rather than spraying pesticides that threaten farmers' health" (*Le Monde*, 2014).

33 English translation: *bugs*, *God-like beasts*.

34 English translation: "Against *little pests*, gardeners use other *little pests*. *[...] Long live munchers*" (*L'Est Républicain*, April 2014).

35 English translation: "The Koppert company, a world leader in integrated biological protection and official supplier of *God's given munchers*" (*La Voix du Nord*, March 2012).

107) "Elle distribue gratuitement des larves de coccinelles et de chrysopes, au Jardin des Plantes, pour *croquer les pucerons de manière écolo*"[37] (*Ouest-France*, 2016).

108) "Small vials of thousands of these barely visible predatory mites can be bought online and simply sprinkled over the leaves of affected plants, *just like pepper*" (*The Guardian*, 2018).

A specific lexical field related to gastronomy is present in both languages, with lexical units such as *dévorer, bâfrer, feed, devour, munch/chew (on), buffet, appetite*, etc. The following extract makes a connection between chillies (*piments*) and a comfortable habitat for aphids (*pucerons*) and whiteflies (*aleurodes*):

109) "Whitefly and aphids are fond of indoor chillies. *Such a comfy life: central heating, no predators and regularly watered, tender plants*" (*The Guardian*, 2015).

We can also illustrate this phenomenon of analogical reasoning with examples from the world of popular culture, in particular scenes from films (science fiction or horror):

110) "La dame pond dans le puceron. La larve *dévore le malheureux de l'intérieur*. Ça ne vous rappelle pas un film de science-fiction?"[38] (*La Voix du Nord*, 2014).

111) "Dans ce huis clos de plastique, se joue le *remake entomologique du film Alien*"[39] (*Ouest-France*, 2018).

112) "There is a *ghastly* fascination in peering through a magnifying glass at a purposeful ladybird *cramming mealybugs into its mouth* with the delight of a five-year-old at a birthday tea" (*The Independent*, 2013).

---

36 English translation: *to crunch aphids*.

37 English translation: "She spreads free ladybug and lacewing larvae, at the Jardin des Plantes, *to crunch the aphids in an eco-friendly way*" (*Ouest-France*, 2016).

38 English translation: "The ladybug lays eggs in the aphid. The larva *devours the unfortunate victim from the inside*. Doesn't that remind you of a certain science fiction movie?" (*La Voix du Nord*, 2014).

39 English translation: "In this plastic enclosure, the entomological *remake of the film Alien is played out*" (*Ouest-France*, 2018).

The lexicon related to such types of films is particularly rich, both in French (*larve*, *dévorer*, *remake*, *film*, etc.) and in English: *ugly larvae*, *gobble them up from within*, *a horror story*, *eating the ladybird from the inside out*, etc.

There are also combinations of analogies. In the following example, the notion of appetite is linked to that of video game (*Game 7*), which is itself associated with bugs doing battle in a sci-fi film:

> 113) "There's a lot riding on the *wasps' appetites*. "This is *Game 7* for these species," said Troy Weldy, director of ecological management for New York's Nature Conservancy. If the idea of *bugs doing battle* sounds vaguely like the premise of a low-budget *sci-fi film*" (*The Wall Street Journal*, 2011).

Lastly, two other examples of fictional scenes show the recurrence of these discursive strategies:

> 114) "A *nematode army* will seek out that target (a grub, slug or some such), *entering its body* and releasing bacteria that cause it to stop feeding and die" (*The Daily Telegraph*, 2016).

> 115) "[…] *Squirming* nematodes in the process of infecting, *devouring and breeding* within slugs in various stages of *gelatinous disintegration*" (*The Daily Telegraph*, 2016).

In these two examples, the auxiliary (*nematode*(*s*)) is described in the way it feeds and develops using linguistic forms from a fictional scenario (*entering its body*, *devouring and breeding*, *squirming*, *gelatinous disintegration*).

The use of analogies has often been analyzed in language for specific purposes (see Oliveira 2009; Condamines et al. 2012), as it provides a simplified access to references. Media discourse also exploits this discursive strategy (e.g. Calsamiglia and Van Dijk 2004; Williams Camus 2009), making it possible to combine a potentially complex situation with another image that is closer to the layman's daily life, therefore rendering it more meaningful. This is a classic simplification strategy in media discourse, for narrative or rhetorical purposes.

The analysis of textual data presented in this section has brought linguistic phenomena to light which, while presenting advantages from a discursive point of view, nevertheless also creates a series of cognitive issues. We present them in detail in Chapter 7, which kicks off Part 3 of our book.

PART 3

Challenges in the Mediation of Agroecological Issues

Introduction to Part 3

Based on the results presented in Chapters 5 and 6, we have produced a summary of the terminological and cognitive issues related to press discourse on agroecology. In Chapter 7, both naming issues (resting on the difference between LSP and LGP) will be addressed, alongside framing issues related to cognitive biases induced by simplification strategies. Chapter 8 will focus on the search for new ideas in order to better communicate on environmental issues, in particular when addressing the lay citizen.

7

Terminological and Cognitive Issues

Simplicity is the ultimate sophistication[1].

Quote from Leonardo da Vinci

The first category of terminological issues identified during analysis of the two corpora is related to the numerous designations of control methods and products. Indeed, analysis of media discourse reveals a series of terms that can be substituted for the main terms in French and English (French: *lutte biologique*, *lutte verte*, etc.; English: *biocontrol*, *organic control*, etc.). Denominative variation has been attested in scientific discourse, as fulfilling different functions (see section 2.2.1). This phenomenon is often more pronounced in media discourse, on the basis of the data analyzed in Part 2. Here, we present the issues related to these important competitive designations in media discourse.

7.1. Designations and specialization

Through the analysis of textual data, we have seen that the choice of predicate or modifier can depend on functional strategies. This first approach is related to communicative registers and relies on the use of non-specialized designations, based on LGP communication. The second approach is a discursive strategy, and involves using variation to focus on non-redundancy. We have also observed interlinguistic variation, which originates from contact between languages and produces variants, to ensure proximity between linguistic (and cultural) communities. Lastly, we observed cognitive variation, which emerges from the differences in conceptualization of the same notion.

1 Original quote: "*La semplicità è la suprema forma di sofisticazione.*"

Tables 5.1–5.3 highlight terminological variation that emanates from the propensity of language for lexical composition. This important combinatorial potential between naming elements leads to terminological growth.

Thus, as the subject develops in the media, new potential combinations are being used, either through a loan translation (for instance, *contrôle biologique* in French, based on "biological control") or on the basis of morphosemantic equivalence (such as *régulation écologique, contrôle naturel, défense biologique; natural warfare or organic defence*). These numerous and disparate designations illustrate an important trend in media discourse, even though this field is still minutely represented in the press.

These practices lead to questioning the impact of this potential for combination, particularly because the lexical units used in the formation of these terms (e.g. *protection, defense, control, natural*, etc.) are also used in other specialized fields. Special attention should be given to this shift between domains.

7.1.1. *Semantic shift between areas of expertise*

A semantic shift is a change in the meaning of a lexical unit, which will be discussed here in relation to another field. Let us take the example of the lexical field linking the two French terms *contrôle* and *régulation* (see section 5.1.1). On the basis of semantic, in particular synonymous relations, other predicative forms are likely to be included in the same lexical field, such as the French terms *surveillance* or *maîtrise*. However, the term *surveillance biologique*, which can be related to the agronomic domain (see section 3.1.1), may also refer to another concept in the medical field[2]. In addition, terms such as *protection biologique* can also be used in the fields of nuclear energy and dermatology[3], and terms such as *bioprotection* and *biovigilance* in the agricultural field[4]. A semantic shift is a natural linguistic process and could be considered as a necessary process responding to the principle of economy of language (for instance, the French term *déficit*, which is

2 "Biological exposure monitoring consists of measuring the level of a biological indicator of exposure to a chemical substance in the blood, urine, hair or exhaled air of exposed workers." See: https://www.inrs.fr/risques/mesure-expositions-agents-chimiques-biologiques/surveillance-biologique-exposition.html [Accessed March 10, 2022].

3 Nuclear: https://www.asn.fr/lexique/P/Protection-biologique. Dermatology: http://nuxeo.edel.univ-poitiers.fr/nuxeo/site/esupversions/0d0fa281-613f-4d09-ba35-2e1b5954dfc9/ [Accessed March 10, 2022].

4 Bioprotection in agriculture: "seed bioprotection", which is one of the methods of protection against diseases through the genetic manipulation of seeds. Biomonitoring: https://draaf.grand-est.agriculture.gouv.fr/Suivi-des-Effets-Non-Intentionnels,876 [Accessed March 10, 2022].

used in the medical field alongside *déficit immunitaire*, and in finance with *déficit budgétaire*). This recycling of terms, which makes it easier for readers to remember the meaning of a lexical item, is therefore a useful process, notably within the same genre or register. However, in press discourse (which is multidisciplinary by its very nature), lexical combinations and shifts between domains can become a potential source of ambiguity.

In the same manner, we can question its impact when it comes to sensitive or controversial subjects. Let us take the example of the lexical field of conflict that was previously identified (*guerre biologique, biological warfare*, etc.), in reference to biological control methods. The term *biological warfare* also appears in a completely different context (in our corpus), relating to the military program developed by North Korea, and more specifically, the hypothetical development of biological weapons (biological neurotoxic agents):

> 116) "In a country that is famously secretive, it is perhaps the most carefully guarded secret of all. North Korea consistently denies having a *biological warfare* program of any kind"[5] (*The Washington Post*, 2017).

In this context of war, the presence of terms such as *bioweapon* or *biological defense* should also be noted; these potentially relate to the combinatory paradigm which was previously identified for biological control, but which instead refers to a military program. This ambiguity is even more significant when the two subjects intertwine. In the following corpus extract, for example, the journalist points to the fact that the industry that makes biopesticides (according to the Korean media) actually corresponds to factories that make biological weapons (according to American analysts):

> 117) "State-run news media described the institute as a factory for making *biological pesticides* – mainly, live bacteria that can kill the worms and caterpillars that threaten North Korea's cabbage crop. But to U.S. analysts studying the video, the images provided an unexpected jolt: On display inside the *military-run facility* were rooms jammed with *expensive equipment*" (*The Washington Post*, 2017).

In this lexical proliferation of predicates and qualifiers, both the journalist and the reader of the press are faced with terminological and cognitive issues. If mediation has to simplify scientific notions, we speculate on its capacity to do so through naming variation, based on common lexical units. This analysis can be related to the phenomenon described by Stosic and Rebeyrolles (2021), which deals

5 This article was collected because it included the term *biopesticide* (see extract 117).

with unspecific nouns in French, such as *chose, objet, idée, problème, personne, principe,* etc.[6]. In the case of biological control, the terms used are also very common lexical units, the difficulty also emanating from the close relationship between the meaning in specialized language and that in general language. This will be discussed in section 7.1.2.

7.1.2. *Determinologization*

The process of determinologization involves a shift in terminology from LSP to LGP, which has an impact on the meaning of terms (e.g. Meyer and Mackintosh 2000; Condamines and Picton 2014; Humbert-Droz 2021; Picton et al. 2021). As noted in section 2.1 – and according to Condamines and Picton – this is most often a phenomenon that neither authors nor readers are aware of[7] (Condamines and Picton 2014, p. 168).

Here, we are interested in the shift of the studied terms (*lutte, biologique, control, organic, parasite,* etc.) into ordinary language, via press discourses. If these terms lose their specialized meaning during this shift, the problem becomes even more complex, as some units are also very common in general language. The interaction between specialized meaning and general meaning is likely to present an interpretative issue for the reader. As previously shown (see section 5.1.1), the reader can rely on a general language definition to construct their own interpretation of the linguistic forms encountered in press articles that deal with a specialized subject. This is potentially what causes difficulty in the interpretation of certain qualifiers (*biologique* in French, for instance, or *organic in English*), which have at least two distinct meanings. It is also, hypothetically, what generates conceptual metaphors in certain fields (e.g. the term *guerre (war)*, which emanates from the lexical field of *lutte* (*fight*) or *conflit* (*conflict*) in everyday language), even though these control mechanisms have a more moderate principle (limiting and regulating pest numbers, rather than eradicating them). However, this question also arises for other terms used in a particular field of specialization, such as *parasite* (*parasite*) or *ravageur (pest)*.

The form *parasite* is used in specialized discourse to designate the (useful) insect, as shown in section 5.2.1.1. It is, however, also used in its common meaning of "superfluous and annoying", and thus in reference to the target insect (in both languages):

---

6 English translation: *thing, object, idea, problem, person, principle,* etc.

7 The authors point out that this principle is opposed to the popularization of science, which implies a deliberate action to make knowledge known to the general public.

118) "La protection biologique intégrée est une technique de lutte contre les *insectes parasites*"[8] (*La Nouvelle République du Centre*, 2013).

119) "Some are a bloody nuisance to us [...]. The book steers clear of these *parasites*" (*The Washington Post*, 2018).

This can be difficult to understand in some extracts. The following looks at the specific case of the chestnut gall wasp, which is considered a pest for this tree, and which has a parasite (in its specialized sense of being a useful insect), as expressed here:

120) "Le retour à l'équilibre entre le *cynips et son parasite* n'est pas attendu avant huit ans"[9] (*Le Parisien*, 2010).

However, other extracts refer to the gall wasp as the "parasite" of the chestnut tree, using its common meaning of "superfluous and annoying":

121) "Une petite guêpe, un *micro-hyménoptère ravageur, un parasite* majeur du châtaignier sévit sur le territoire. Le cynips, [...]"[10] (*La Montagne*, 2014).

The gall wasp's own *parasite*, *Torymus sinensis*, acts as a parasitoid, therefore making the situation difficult to understand in some contexts:

122) "Les larves issues de ces œufs se nourrissent alors des larves de cynips empêchant ainsi *l'émergence* du *parasite*. Ce *parasitoïde* n'est pas naturellement présent actuellement dans la nature. Il faut *l'introduire*"[11] (*La Montagne*, 2014).

In this case, the gall wasp is described as a *parasite* of the chestnut tree that is being controlled, whereas from the specialized discourse perspective, it is *Torymus sinensis* that is seen as the gall wasp *parasite* (and even a *parasitoid*)[12]. With little

8 English translation: "Integrated biological protection is a technique for controlling *parasitic insects*" (*The New Republic of the Center*, 2013).

9 English translation: "The return to balance between the gall wasp and its parasite is not expected for eight years" (*Le Parisien*, 2010).

10 English translation: "A small wasp, a devastating *micro-hymenoptera pest, a major parasite* of the chestnut tree is rampant in the territory. The gall wasp, [...]" (*La Montagne*, 2014).

11 English translation: "The larvae from these eggs then feed on the gall wasp larvae, thus preventing the *emergence* of the *parasite*. This *parasitoid* is not naturally present in nature. It must be *introduced*" (*La Montagne*, 2014).

12 See the glossary at the end of the book or the online Vegephyl dictionary for the difference between the two terms (https://www.vegephyl.fr/publications/bases-de-donnees/base-repertoire-terminologie/).

specialized knowledge, the lexicon in use and the co-text will be particularly useful for interpretation. Here, in example 122, the noun *émergence* (*emergence*) or the verb *introduire* (*introduce*) may allow the reader to better understand the different roles in this control process; however, the discourse remains undeniably complex from a terminological and cognitive point of view, especially for a non-specialist.

In the English corpus, the reader can also rely on contextual lexical elements to reconstruct the discourse (here, using the verb *to introduce*):

> 123) "The best option for whitefly is to *introduce* Encarsia, a tiny *parasitic biological control*" (*The Guardian*, 2015).

Lastly, we note the use of the French term "prédateurs des plantes" (plant predator) in media discourse, which is not used in specialized terminology, but makes it possible to explain the process:

> 124) "Cela impose d'autres pratiques et d'être sans cesse à la recherche de l'équilibre entre les *prédateurs des plantes*, et leur ennemi naturel qu'il soit animal ou biologique"[13] (*La Nouvelle République du Centre*, 2011).

In specialized terminology, *predator* only refers to the auxiliary, and when plants are the target, we speak of (crop) *pests* (see glossary). Even though the co-text may help the reader to understand the situation, we agree with Mortureux's analysis that the expected simplification of popularization seems, at times, to be compromised by the complexity of the semantic relations in these discourses (Mortureux 1994, p. 223).

In any case, interpretation requires a certain capacity for attention from the reader, as well as a significant cognitive effort (de Cheveigné 1997) in order to identify the precise role of each insect in the process.

A third issue, linked to naming, raises other interpretative questions related to connotations. These are presented below.

7.1.3. *Designations and connotations*

The use of suffixes or prefixes from specialized terminology raises a number of questions. Here, the case of the suffix *-cide* and the prefix *phyto-* will be discussed.

13 English translation: "This imposes other practices and to be constantly in search of the balance between the *plant predators* and their natural enemy, whether it be animal or biological" (*La Nouvelle République du Centre*, 2011).

7.1.3.1. *The suffix -cide*

The suffix *-cide* comes from the Latin *cædere*, meaning "to kill". Our double corpus contains a large paradigm of terms built on this suffix (*pesticide, insecticide, fungicide, parasiticide, biopesticide*, etc.). The meaning of this relatively widespread suffix is somewhat ambiguous, even in ordinary language. Nevertheless, as Crump et al. (1999) and Eilenberg et al. (2001) point out, its meaning is connotated:

> The use of the suffix "-cide" tends to highlight an association with pesticides. Whether this is a negative association in the mind of environmentalists, or an expectation of rapid kill in the mind of farmers, it is clearly an association which should be avoided if inundation biological control is to achieve wide acceptance (Eilenberg et al. 2001, p. 396).

According to these authors, the social and professional acceptability of the processes involved may depend on these terminological questions, especially if the (specialized) terms used are not likely to convey the nature of these processes; in this case, it is achieved by its reference to a target mortality (rapid and direct), instead of a regulation process (of pest numbers).

In press discourse, the problem is exacerbated, as the reader may not necessarily have the prerequisites to accurately interpret the extracts. It is even more complex as the terms using this suffix refer to distinct types of products with different strategies. This point is demonstrated with two corpus extracts, including the French terms *insecticide* and *fongicide (fungicide)*. In the first example, they refer to chemically synthesized plant protection products, the aim of which is to get rid of the pests without seeking to regulate their numbers:

> 125) "Tous les produits ne sont pas en cause. On parle d'antilimace, de *fongicides, d'insecticides*, ou de désherbants au glyphosate, les plus communément utilisés"[14] (*Le Figaro*, 2015).

This association between the suffix *-cide* and the aim of rapid mortality is confirmed by the co-text of these elements, associating glyphosate with these products "*les plus communément utilisés*" ("the most commonly used"). In contrast, the same two terms below (*insecticides* and *fongicides*) can also refer to natural products, made "from plants":

---

14 English translation: "Not all products are included. We are talking about slug pellets, *fungicides, insecticides*, or glyphosate weed killers, which are the most commonly used" (*Le Figaro*, 2015).

126) "On peut en effet, à partir de plantes, concocter des engrais, des *insecticides et des fongicides* très utiles aux plantations qui vont bientôt débuter"[15] (*La Montagne*, 2017).

The objective of these "natural" products is to regulate the number of pests and not to destroy them.

An additional semantic issue arises from these first elements: the fact that the French term *pesticide*, also built on this suffix, is commonly considered to be an equivalent of the term *produit phytosanitaire* or *produit phytopharmaceutique* (*phytosanitary product* or *phytopharmaceutical product*). Let us examine the meaning of this other prefix *phyto-*, as well as its uses in context.

7.1.3.2. *The prefix phyto-*

In etymological terms, *phyto-* means "plant", while the adjective *sanitary* refers to health. The term *phytosanitary* thus refers to the health of plants, and is generally used as an equivalent to the adjective *phytopharmaceutical*, which characterizes plant medication. Therefore, there is no negative connotation attached to these lexical forms a priori. Some extracts recall the relatively neutral semantics of these units:

127) "Ensuite nous faisons une stratégie de *phyto-protection* sans agents chimiques"[16] (*L'Humanité*, 2014).

128) "Ces *méthodes phytosanitaires* à faible risque pour la santé et l'environnement [...]"[17] (*La Croix*, 2014).

Nevertheless, given these non-connotative uses of terms based on this prefix (*phyto-*), there are many other interpretations in media discourse, notably when it is used in the term "phytosanitary product" (or "phytopharmaceutical product"):

129) "Les *produits phytosanitaires* sont communément appelés pesticides chimiques. Ils comprennent les insecticides, les fongicides, les herbicides, les parasiticides"[18] (*La Nouvelle République du Centre*, 2016).

---

15 English translation: "We can indeed make fertilizers, *insecticides* and *fungicides* from plants that are very useful for the planting that will soon begin" (*La Montagne*, 2017).

16 English translation: "Then, we use a *phyto-protection* strategy without chemical agents" (*L'Humanité*, 2014).

17 English translation: "These *phytosanitary methods* with low risk for health and the environment [...]" (*La Croix*, 2014).

Phytosanitary products here commonly refer to pesticides; in other words, chemical products, which can have a negative connotation. This adjective (*phytosanitary*) is also used in a number of other terminological combinations to refer to chemical methods:

> 130) "Le scientifique messin était un farouche opposant à l'usage de la *chimie phytosanitaire*"[19] (*L'Est Républicain*, 2017).

> 131) "Et même si pour l'heure, le *monde phytosanitaire* ne perd pas de terrain, il n'en subit pas moins ces pressions"[20] (*La Tribune*, 2016).

These methods and products thus produce a negative connotation in public and collective imagination, to which alternative solutions are being sought.

Other discursive phenomena maintain this implicit meaning. Indeed, since the Ecophyto plan (2008), various regulations (e.g. the Labbé law (2014) or the law on energy transition (2015)) have imposed a progressive decrease in the use of phytosanitary products in plant protection. The phrase "*objectif zéro phyto*" ("zero phyto objective") has been coined in the legal framework to encourage these practices. However, this phrase, which can be likened to a "formula" (Krieg-Planque 2009), raises questions insofar as it is not based on the etymology of "phyto", but rather on its discursive (and cultural) uses, which have depreciatory connotations (referring to a chemical product). Thus, this complexity remains with biocontrol products, which are defined in the LAAF law as "phytopharmaceutical products", and which are "natural" products recommended by the aforementioned legal texts. Moreover, these are subject to similar regulations as those for conventional phytopharmaceutical products, as the following extract recalls:

> 132) "L'ANSES (Agence nationale de sécurité sanitaire) délivre les autorisations de mise sur le marché (AMM) des *produits phytosanitaires, chimiques ou biologiques*, notamment pour l'agriculture"[21] (*Le Figaro*, 2018).

---

18 English translation: "*Phytosanitary products* are commonly referred to as chemical pesticides. They include insecticides, fungicides, herbicides and parasiticides" (*The New Republic of the Center*, 2016).

19 English translation: "The scientist from Metz was a fierce critic of the use of *phytosanitary chemistry*" (*L'Est Républicain*, 2017).

20 English translation: "And even though, for the time being, the *phytosanitary world* is not losing ground, it is nonetheless subject to these pressures" (*La Tribune*, 2016).

21 English translation: "The ANSES (French National Health Safety Agency) issues marketing authorizations (AMM) for *phytosanitary, chemical* or *biological* products, particularly for agriculture" (*Le Figaro*, 2018).

	French language	English language
Generic designation (origin/composition)	– *Pesticide, pesticide chimique, pesticide chimique de synthèse* – *Produit chimique, produit de synthèse chimique* – *Produit phytosanitaire (phyto sanitaire, phyto-sanitaire), produit phytopharmaceutique (phyto pharmaceutique, phyto-pharmaceutique), produit phytopharmaceutique de synthèse, produit phyto, phyto phytosanitaire* – *Protection chimique* – *Traitement chimique, traitement phytosanitaire, traitement phyto* – *Chimie, chimie de synthèse, chimie phytosanitaire* – *Agent chimique, molécule chimique, arme chimique, arsenal chimique, solution chimique, substance chimique, substance de synthèse* – *Filière phytosanitaire* – *Monde phytosanitaire* – *Fongicide (chimique)* – *Insecticide (chimique)* – *Produit toxique, substance toxique, molécule toxique, produit neurotoxique*	– Chemical(s) – Chemical pesticide – Chemical product – Chemical spray – Chemical treatment – Chemical control – Chemical methods – Chemical variant – Chemical synthesis – Chemical weapon – Chemical warfare – Pesticide – Agricultural pesticide – Synthetic biocide – Synthetic compounds – Synthetic ingredients – Neonicotinoid treatment – Neonicotinoid pesticides – Insecticide – Fungicide – Biocide – Abrasive – Aggressive – Caustic
Impact on biodiversity	– *Pesticide nocif, produit nocif, traitement nocif, molécule nocive, pratique nocive, matières actives nocives* – *Produit dangereux, insecticide dangereux, traitement dangereux, molécules dangereuses* – *Produit polluant, traitement chimique polluant, phytosanitaire polluant*	– Dangerous – Deadly – Detrimental (to) – Environmentally harmful – Harmful – Hazardous – Risky – Toxic – Unsafe (all are followed by the terms chemical/pesticide or product, here and below)
Product specificity	– *Produit à large spectre/à spectre large* – *Pesticide tueur*	– Broad-spectrum – Chemical kills – Kill-everything – Quick-fix
Miscellaneous	– *Granulés bleus* – *Tambouille chimique*	– Nasty – Unpopular

Table 7.1. *Competitive designations of chemical products and methods*

To illustrate this complexity, the extended combinatory paradigm for "chemical" methods is presented in Table 7.1; this paradigm of terms is opposed to the designation paradigm for "biological" methods as shown in the tables of Chapter 5.

We wish to show that, through naming variation, discourse objects are constructed and modified, while naturally moving away from the terminological elements of specialized discourse.

As Moirand points out when analyzing sensitive subjects (e.g. GMOs or bovine spongiform encephalopathy), different semantic relations are involved in discourse (synonymy, hypernymy and hyponymy, metonymy and metaphorization), thus creating an important series of co-references in the representation of the discourse object through the text. The author also points out that among all of the forms produced, "images constructed by association or analogy do not show any trace of scientificity"[22] (Moirand 2007, p. 25). This is also the case in our study, with designations such as *pesticide tueur (*killer pesticide*), granulés bleus (*blue pellets*), or* or *quick-fix chemical.*

We now move on to another issue related to naming, which pertains to the discursive equivalences between terms.

7.1.4. *Name equivalents in discourse*

We have previously looked at the complex relationship between the terms *lutte biologique* (*biological control*) and *biocontrôle (biocontrol*) (see Chapter 4). This extract, which was published several months after the LAAF law, illustrates this:

> 133) "Elles ont vu l'utilisation des pesticides baisser en moyenne de 12 % en 2013, grâce à la rotation des cultures, la diversification variétale ou le recours au *biocontrôle* [la *lutte biologique*]"[23] (*Libération*, 2015).

The term *lutte biologique* (*biological control*) in square brackets is juxtaposed with *biocontrôle*, yet the article does not explain the difference between these two terms. While related by the fact that *lutte biologique* is part of the definition of "biocontrôle", it is only a part of it (living organisms). The reader's interpretation of

22 Original French version: "*Images construites par association ou analogie ne présentent pas de trace de scientificité*" (Moirand 2007, p. 25).

23 English translation: "They saw pesticide use drop by an average of 12% in 2013, thanks to crop rotation, varietal diversification or the use of *biocontrol [biological control]*" (*Libération*, 2015).

this extract depends on their own expertise; if they have some form of general knowledge about this field, they may deduce that, among natural mechanisms for plant protection, biological control is the preferred one here. If they are less knowledgeable, they are likely to conclude that these are synonyms. Lastly – and this is probably the journalist's objective – they are likely to understand the recently coined term *biocontrôle* by linking it to the potentially better-identified term *lutte biologique,* without going into too much technicality in explaining their difference.

The example below shows a similar situation, with the assimilation of biocontrol methods (natural substances) to those of biological control:

> 134) "Cette sorte de petite guêpe est l'un de ces macro-organismes prédateurs qui s'inscrivent avec les micro-organismes pathogènes, les médiateurs chimiques ou les *substances naturelles* dans l'arsenal déployé par la *lutte biologique* pour se débarrasser des ravageurs des cultures"[24] (*L'Est Républicain*, 2016).

While natural substances are an integral part of "biocontrôle", they are not part of the original processes of biocontrol (the latter involving living organisms, macro- and micro-organisms). However, this comparison allows the journalist to illustrate the potential control processes, and makes it possible for the reader to identify different methods for controlling pests using natural methods.

The following explores a final issue related to designations, which is linked to the status (whether living or not) of control agents and products.

7.1.5. *Agent and product status*

In Chapter 4, it was highlighted that specialized definitions of terms already incorporate a form of instability, which concern the status of control agents. The situation is made more complex as the LAAF definition refers to living organisms (micro-organisms) under the term "products" (see section 4.1.1).

Media discourse is also faced with the difficulty of defining discourse objects. In the following extract, for example, the journalist relies on a specific lexicon to show that it refers to living organisms (the verb *lâcher* in French, *release* in English):

---

24 English translation: "This kind of small wasp is one of those predatory macro-organisms that correspond to pathogenic micro-organisms, chemical mediators or *natural substances* in the arsenal deployed by *biological control* to get rid of crop pests" (*L'Est Républicain*, 2016).

135) "Si ça dépasse, on *lâche* la juste quantité d'*auxiliaires*, qui vont se nourrir des parasites"[25] (*La Nouvelle République du Centre*, 2014).

136) "Vandenberg, a scientist with the U.S. Department of Agriculture, would *release* the *insects* – two species of wasp" (*The Wall Street Journal*, 2011).

However, auxiliaries (insects) are also described as "products", as in the following example:

137) "L'utilisation de produits phytosanitaires y est donc interdite, mais il est encore possible d'utiliser des *produits de bio-contrôle – comme des insectes* ou des substances naturelles"[26] (*Ouest-France*, 2017).

When it comes to "inert" products (such as natural substances), the introductory predicate is potentially different, and the notion of *application* can be found:

138) "Une agriculture de précision pour éviter la perte au moment de *l'application* si on cible la bonne dose sur la culture"[27] (*Le Figaro*, 2017).

139) "These products are contact insecticides and must be *applied* on a weekly basis until no more spider mites can be found" (*The Sun*, 2015).

Nevertheless, for some living organisms (which, in this example, are mites), we also find predicates that are usually associated with products, such as *diffusion* (*spread*), *saupoudrage (sprinkling)*, or *spray*:

140) "Ainsi, la mouche blanche, qui menace les poinsettias, va être combattue avec ses *prédateurs naturels*, les acariens, soit *saupoudrés* sur les plants, soit *diffusés* dans des sachets"[28] (*Ouest-France*, 2011).

141) "We *sprayed* on the T-warriors, following the routine we had used for crane-flies and Sportnem" (*The Financial Times*, 2017).

---

25 English translation: "If it goes over, we release the right number of beneficials, which will feed on the pests" (*La Nouvelle République du Centre*, 2014).

26 English translation: "The use of phytosanitary products is therefore prohibited, but it is still possible to use biocontrol products – like insects or natural substances" (*Ouest-France*, 2017).

27 English translation: "Precision agriculture to avoid loss at *application* if the right dose is targeted to the crop" (*Le Figaro*, 2017).

28 English translation: "Thus, the whitefly, which threatens poinsettias, will be fought using its *natural predators*, the mites, either *sprinkled* on the plants or *spread* out in sachets" (*Ouest-France*, 2011).

According to Eilenberg et al. (2001), this is originally a matter of perspective: "Entomologists speak often of release, while plant pathologists use the term *application.*" Their analysis, which concerns LSP, is not necessarily applicable to LGP and press discourse.

The section below continues the presentation of issues arising from textual analysis by focusing on the characterization of methods (through extended textual structures), and their discursive interpretation. We first begin by introducing the issues associated with this Manichean approach to discourse objects.

7.2. Binarism and cognitive bias

In section 6.1.2, we described a form of binary framing that characterizes the presentation of methods, agents and products, which are partly biological and partly chemical. As observed, the main meaning of the qualifier is not always a determining factor for its interpretation, as each term can be substituted in discourse (e.g. *biologique/vert (biological/green)* or *chimique/classique (chemical/classic)*). What is significant, however, is the opposition between the two sub-categories of products and methods, some of which are considered as "natural" (biological), and others as "synthetic" (chemical). Thus, with little knowledge of the field (but with socially shared knowledge, or a kind of common sense), the reader can easily classify the product or method on either side of a binary framing, which can be positive versus negative. These rhetorical strategies have strengths and weaknesses, which will be detailed below.

7.2.1. *Strengths and weaknesses of binary reasoning*

As Geeraerts et al. underlined in their extensive study of clothing terminology, "the major differences that we have found appear to involve pragmatic, situational factors rather than sociolinguistic or geographical ones" (Geeraerts et al. 1994: 153). The pragmatic factors are consistent with Sinclair's findings on "semantic prosody", who speaks of an "attitudinal and pragmatic meaning", with a potentially negative connotation associated with a lexical form:

> The selection of the item is controlled by the prosody, because the whole point of expressing oneself in this way is to pre-evaluate the actions, which would otherwise be positively evaluated by the reader/listener (Sinclair 2000, p. 201).

We can also refer to the notion of axiology (in French, *axiologisation*). The assumption of a depreciative value of a linguistic form can be applied to the form "chemical". Since the lexical units (biological and chemical) are repeatedly opposed on the positivity/negativity scale, the purpose of each unit seems to stem from one of the two perspectives. Indeed, as Bednarek and Caple (2014) and Molek-Kozakowska (2017) have shown, this oscillation between negativity and positivity is an important stylistic pattern in science popularization discourses, with a view to reporting opposing perceptions. For example, in the environmental field, Molek-Kozakowska (2017) attests to the sense of security provided through positively presented knowledge, explanation, warnings or remedies. Conversely, the discourse may refer to an emotion (or attitude) that is conveyed in a negative way, or lexical items that refer to undesirable situations and actions.

The negative connotation and emotion thus transmitted is based on a socially determined rhetoric (or belief) that pesticides are artificially produced, and are therefore polluting and toxic. These opposing discursive processes can certainly simplify the argument by triggering emotional reactions from the public. As Grize pointed out:

> In our daily activities, feelings are as important as reason and an argument cannot be settled by addressing intelligence alone, it must also speak to an individual emotionally. It is not enough to just to just move people emotionally to communicate ideas and make them know, it is also necessary to make them see what happens through the use of discourse objects[29] (Grize 2004, p. 42).

These principles are based on the fact that prior cultural references (in French, *préconstruit*) that allow the reader to interpret discourses are not neutral: "Each of them is situated on a euphoric-dysphoric axis that belongs to the reader"[30] (Grize 2004, p. 42).

This binary textual mechanism serves different purposes. It makes complex concepts more accessible, and provides the reader with simple and immediate means of accessing knowledge on a subject. Indeed, as Rastier points out, "binarism is attractive because it is easy to understand. Experimental work carried out over the last century on associations and the organization of the mental lexicon has always

29 Original French version: "*Dans nos activités de tous les jours, les sentiments importent autant que la raison et une argumentation ne peut se contenter de s'adresser à la seule intelligence, elle doit encore émouvoir. Il ne suffit pas pour émouvoir de communiquer des idées donc de faire savoir, encore faut-il faire voir ce qui se réalise par l'usage des figures du discours*" (Grize 2004, p. 42).

30 Original French version: "*Chacun d'eux se situe sur un axe euphorique-dysphorique qui appartient au destinataire*" (Grize 2004, p. 42).

confirmed that the antonym comes to mind in the shortest possible time"[31]. Our observations have shown the importance of this relation in our corpus, which, according to this author, "requires little effort" (Rastier 2011, p. 122).

However, a set of issues linked to this binarism have been identified (biological is positive and chemical is negative). How can this Manichean opposition be a source of interpretative complexity, or even a hurdle in communication? The answer to this question is illustrated here by a series of related queries, in connection with plant protection issues, that the general public is likely to ponder when reading the press.

7.2.1.1. *Should all chemical pesticides be banned?*

Let us return to the definition of the adjective *chemical*, which in general language refers to "the structure of substances". Thus, from the reference definition perspective, there is no negative connotation associated with this adjective, despite the depreciative value observed in discursive semantics[32]. However, should all "chemical" products be banned?

It is important to note that biocontrol products are produced by scientific processes that use chemistry to mimic mechanisms found in nature. The molecule produced is therefore similar to a natural molecule, but with varying degrees of modification. Biocontrol products are therefore largely derived from chemical synthesis, as testified by the definition of *biocontrôle* in the LAAF law; it uses the term *produit phytopharmaceutique* (*phytopharmaceutical product*) to refer to biocontrol products, just like for any other conventional pesticide.

Thus, a biocontrol product such as pheromones is defined as a "chemical mediator", based on a substance that can be compared to hormones[33] and emitted by most animals – as well as by certain plants. It is therefore a chemical substance of animal origin, and therefore natural. Following a terminological logic, this product could be called a "chemical (bio)pesticide". It would also correspond to the

31 Original French version: "*Le binarisme séduit parce qu'il est facile à comprendre. Les travaux expérimentaux menés depuis un siècle sur les associations et l'organisation du lexique mental ont toujours confirmé que l'antonyme vient à l'esprit dans le temps le plus court*" (Rastier 2011).

32 This principle was already recalled in 2014 in: Chemical is not a dirty word. *New Scientist* [Online]. Available at: https://www.newscientist.com/article/mg22429973-100-chemical-is-not-a-dirty-word/ [Accessed January 17, 2024].

33 See: https://www.lesartisansduvegetal.com/conseils-videos-les-artisans-du-vegetal/pheromone-kairomone-au-jardin-comment-cela-marche-6063.html [Accessed February 25, 2022].

definition of a "natural (bio)pesticide". Lastly – and ironically – it could be called a "natural chemical". In any case, the simplifying and recurrent discursive strategy that emphasizes the absence of chemistry in these processes (no chemistry, no chemical, etc.) is not relevant from a scientific perspective.

These observations show that the prevailing binary framing (opposing what is biological to what is chemical) is adequate for discursive and explanatory purposes, but inadequate for the transmission of a minimum of cognitive elements. The negative connotation attached to chemically synthesized pesticides does not a priori convey the complementarity of processes (and even of disciplines). Moreover, the terms *biochimie* or *biochemistry*, which are widely used in the scientific community, correspond to the following definition: "scientific discipline that deals with the chemistry of living beings". These terms (or the adjectives *biochimique* and *biochemical*, which are absent from our corpus) attest to the fact that biology and chemistry are disciplines that are complementary and not opposed.

We now move on to a second line of questioning that is brought about by the generally positive way in which biological methods are presented in the media corpus.

7.2.1.2. *Should biopesticides be systematically encouraged?*

The fact that biopesticides are less harmful to the environment than their "chemical" counterparts is a prevailing view throughout press discourses. However, this perspective can also appear to be one-sided and lacking in nuance.

In fact, biocontrol products can also be considered harmful to the environment. Some corpus extracts mention the case of fungicide products (e.g. copper-based) and their supposed toxicity:

> 142) "*Copper fungicides* (approved for organic use) can be toxic to the plants they are intended to protect" (*The Wall Street Journal*, 2015).

Other arguments are proposed to show the reader that the reproduction of natural mechanisms can also lead to environmental issues. One argument used in the following example is the pollution involved in these methods (more trips made by tractors, and therefore more carbon dioxide emissions into the atmosphere):

> 143) "*Pest control*, too, often involves sulfur, coppers or biological pesticides, which *require many more tractor trips* to deliver than their conventional-farming equivalents, thus *increasing carbon in the atmosphere*" (*The Wall Street Journal*, 2015).

Linking pollution to biological control methods offers a contrasting perspective to the "classic" (and prevalent) view described above. However, it can establish an interesting counterpart when presenting different issues related to this field.

Another argument that the general public is responsive to is the risk generated by these mechanisms when they disrupt the ecosystem:

> 144) "*Importing tiny insects* to tackle Britain's Japanese knotweed problem *could easily backfire on us*, warns Michael Leapman. *Meddling with nature is always controversial*" (*The Daily Telegraph*, 2010).

Indeed, as previously described, another perspective on these products involves their impact on the ecosystem, particularly over the long-term. Without delving into the principle of potential disruption (we will return to it in Chapter 8), it is important to emphasize that this binary perspective may not explain the full complexity linked to the processes involved.

These framing effects focus on certain angles, while generating numerous "blind spots". In her analysis of how the media presents pesticides, Botero analyzes the lack of information about food products and their production methods. Food, in general, is neglected in these discourses, as is the farmer's voice, or that of associations and citizens aimed at criticizing the harmfulness of these controversial products (Botero 2019, pp. 116–117). As such, we can evoke a difference in terms of framing information, as established by Iyengar (1991), contrasting episodic and thematic framing: episodic framing focuses on specific, one-off events or particular cases to illustrate an issue, while thematic framing places events in a general context, in order to question them (Marty 2015). Episodic framing is therefore more about presenting information in a simple manner, without necessarily requiring an in-depth analysis. Binarism falls under this first framing, with stories that have an immediate impact. Thematic framing corresponds to a more analytical presentation of the issues associated with the subject. The second perspective makes it possible to express the complexity of the processes, expanding the reflection to societal impacts of such methods and products (environmental protection, biotechnologies, health issues, etc.).

Once again, the press has the difficult task of relying on these two types of framing, not only using punctual and immediate impact elements to capture (and entertain) a reader, but also going deeper into the subject (through thematic framing), and running the risk of not "appealing" to a potential reader (see Chapter 8).

Another hurdle is the issue of classicism in plant protection methods, as described previously.

7.2.1.3. *Should chemicals be considered as classic products?*

As seen in section 6.1.2.3, the adjectives *classique*, *habituel*, *traditionnel*, *conventionnel* (*classic*, *usual*, *traditional*, *conventional*) are regularly used as synonyms in discourse, and overwhelmingly refer to the chemical methods that have been used in agriculture since the 20th century. However, in other extracts, these qualifiers take on different meanings. These are briefly presented below.

The term "*méthodes traditionnelles*" ("traditional methods"), for example, is likely referring to older practices. The connotation of the adjective then becomes increasingly positive, as it is more related to "natural" processes:

> 145) "La municipalité a organisé, samedi, un après-midi dédié aux amateurs de jardinage dans l'objectif de les sensibiliser à ne plus utiliser de produits chimiques et de revenir à des *méthodes traditionnelles*"[34] (*La Voix du Nord*, 2012).

The qualifier refers to "ancestral" traditions – methods that are closer to nature, and are therefore synonymous with terms such as *natural* or *ecological* in everyday language.

The same ambiguity exists regarding the meaning of the term *traditional* in the following English extract, which describes a method for controlling aphids:

> 146) "There are many kinds of *traditional methods* to fight off the aphids" (*The Daily Telegraph*, 2012).

Without the co-text or context, it is difficult to interpret this unit, as well as the term *traditional methods*, which is part of the designation paradigm of chemical methods. The rest of the extract is, as such, a determining factor:

> 147) "[...] *Traditional methods* to fight off the aphids, *from flour sprays to cow urine*" (*The Daily Telegraph*, 2012).

Reading the overall extract, any reader understands that *traditional* refers to ancient and "natural" traditions (flour and cow urine), which were used long before the existence of chemical methods. However, its interpretation depends on

34 English translation: "On Saturday, the municipality organized an afternoon dedicated to gardening enthusiasts with the aim of raising awareness of the need to stop using chemicals and to return to *traditional methods*" (*La Voix du Nord*, 2012).

time-based parameters, as well as on the reader's effort (or ability) to place these elements in the context of what they know about the history or methods of plant protection in agriculture.

In the following example, the interpretation of this adjective can also raise questions as to the relation between chemical (pesticides, fungicides) and biocontrol methods (copper, sulfur):

> 148) "Ces techniques s'utilisent rarement seules, plutôt dans le cadre de ce qu'on appelle la "protection intégrée", c'est-à-dire en complément des *traitements traditionnels: pesticides, fongicides, cuivre, soufre*"[35] (*La Croix*, 2016).

In this extract, the French term *traditionnels* (*traditional*) does not specifically refer to chemical methods or to "old and natural" methods, but rather to the original definition of the term, which means "what one is used to". Indeed, copper and sulfur, which are natural substances that do not fall under the category of chemicals, are nevertheless part of plant protection practices, in addition to chemical methods.

In the following example, the adjective *traditionnel* (*traditional*) has yet another meaning, that of the refusal of innovation (linked to the preference for "usual" methods):

> 149) "[…] Le jour où il a remis en cause ses habitudes de culture intensive, délaissé un peu les pots de fongicide et autres herbicides, où il a réécouté la nature, devant affronter souvent les quolibets d'un *milieu* parfois *traditionnel*" […][36] (*Le Progrès*, 2010).

In this case, the qualifier has a potentially negative connotation, but is not synonymous with "conventional" plant protection methods.

Lastly, there is yet another use of the semantically similar qualifier *conventionnel(le)* (*conventional*), built on the model of this combination paradigm:

> 150) "De plus, les produits carnés issus d'animaux nourris avec des plantes GM ont les mêmes caractéristiques que les produits issus

35 English translation: "These techniques are rarely used alone, rather as part of what is called 'integrated protection', that is to say, as a complement to *traditional treatments*: *pesticides, fungicides, copper, sulfur*" (*La Croix*, 2016).

36 English translation: "[…] The day he questioned his practices of intensive cultivation, abandoned fungicide and other herbicides, when he listened to nature again, often having to face the jeers of a sometimes-traditional *environment* […]" (*Le Progrès*, 2010).

> d'animaux nourris avec des aliments issus de *plantes conventionnelles*"[37] (*L'Humanité*, 2016).

The adjective *conventionnelles* (*conventional*), associated with plants here, means "non-modified, non-GMO", and is likened to the "traditional varieties" of plants. It is therefore synonymous with the term *pure*.

When reading these examples, we note a semantic instability in the interpretation of these different terms, which strongly depends on the co-text and the context, and on the nuance brought about by the qualifier. It is therefore difficult to generalize the semantics of the French term *classique* (*classic*) and its equivalents in discourse. Below, four different meanings of qualifiers are summarized, all of which have been identified in the corpus:

a) in accordance with usages, conventions or practices (e.g. *traitement classique, agriculture traditionnelle, conventional pest control*);

b) respecting the environment or close to nature (e.g. *méthodes traditionnelles, traditional methods*);

c) rejecting innovation[38] (e.g. *milieu traditionnel, traditional environment*);

d) non-modified, non-hybrid (as opposed to GMO) (e.g. *plantes conventionnelles, conventional plants*).

The first meaning is the most common in our corpus, and is generally found in reference dictionaries under most entries (*classique, traditionnel, conventionnel, traditional, conventional,* etc.); this refers to practices, without any particular connotation a priori (such as conventional treatments). On the other hand, given that the practices of the 20th-century agriculture concern the massive use of chemical treatments, this specific discursive meaning prevails in many noun terms (even if the example of the adjective *traditionnel* (*traditional*) is semantically ambiguous in the double corpus).

All of these elements show that interpreting an article requires strong focus from the reader in order to correctly understand the meaning of the terms. This is all the more complex as these adjectives implicitly refer to the general issue of time, and

37 English translation: "Moreover, meat products from animals fed with GM plants have the same characteristics as products from animals fed with food from *conventional plants*" (*L'Humanité*, 2016).

38 In a glossary published by INRA in 2002, the authors already indicated this connotation: "It should be noted that, in the current agricultural context, the expression 'traditional agriculture' can be considered as a qualifier for anti-modernization agriculture" [Online]. Available at: https://hal.archives-ouvertes.fr/hal-01202642/file/C45Blouet.pdf [Accessed February 22, 2022].

thus from a perspective that is both diachronic and synchronic, which leads us to a final question.

7.2.1.4. *Should current agriculture be considered as modern?*

Methods evolve and the question of time appears when reading some extracts:

> 151) "Les *méthodes de lutte actuelles* font appel à des traitements chimiques *très polluants et peu* efficaces"[39] (*Ouest-France*, 2010).

In this example, chemical methods are considered as current (and polluting). The chemical model is still the appropriate, dominant model of the last decade. According to this other extract, what is "contemporary" (*contemporain*) is beginning to be opposed to other, older methods:

> 152) "Car l'*agriculture contemporaine* sait fort bien tirer parti des *nouvelles technologies*"[40] (*La Nouvelle République du Centre*, 2017).

New (and current) agriculture is simultaneously described below, as combining modernity and tradition:

> 153) "Cette "*nouvelle agriculture*" est déjà là. Masquée par les nostalgies et les idéologies, elle propose une nouvelle voie très pragmatique qui allie les *techniques les plus modernes* à des *moyens plus traditionnels*"[41] (*La Croix*, 2015).

These different extracts show a change in perspectives in the face of innovative methods in crop protection. We are, indeed, at a pivotal point in the evolution of practices, which is generating this semantic and temporal instability. The potential difficulty is linked to the position of the writer/reader on this time scale, and that of the co-construction of meaning during interpretation. In this respect, let us observe the following extract:

> 154) "Environmental groups, that contend fertilizers and pesticides are used excessively in *modern agriculture*" (*The Wall Street Journal*, 2012).

---

39 English translation: "*Current control methods* use chemical treatments that are highly polluting and not very effective" (*Ouest-France*, 2010).

40 English translation: "*Because contemporary agriculture* can take advantage of *new technologies*" (*La Nouvelle République du Centre*, 2017).

41 English translation: "This '*new agriculture*' is already here. Masked by nostalgia and ideologies, it proposes a new, very pragmatic path that combines the *most modern techniques* with *more traditional means*" (*La Croix*, 2015).

From the perspective of lexical semantics, the term *modern* means "of the time of the speaker or of a relatively recent period". It is given as a synonym for *current* or *contemporary*, and could then be considered as the antonym of the terms *traditional* and *conventional*. However, in this example, "modern agriculture" is used as a discursive equivalent of what was described earlier as "conventional" agriculture; in other words, it refers to the use of chemical methods (*modern* here refers to the post-industrial revolution), during which synthetic products have shown their efficiency on a large scale. Some adjective are thus shown to be "opaque", as their interpretation depends on spatio-temporal criteria (Kleiber 1979, 2008). The press article in question happens to point out the criticism about companies that massively used these products when agriculture began to employ more industrial methods. This apparent discursive paradox, in which modernity is assimilated to traditional methods, can only be resolved by the reader's ability to analyze the context and interpret the meaning of the term; this is based on their personal knowledge of the evolution of plant protection methods.

We can clearly see the difficulties of explanation and interpretation that are linked to these subjective temporal elements, in the face of a heavily complex situation.

The following is a look at the strengths and weaknesses of other strategies for simplifying processes, methods and products in press discourse.

7.2.2. *Strengths and weaknesses of analogical reasoning*

We have already discussed the discursive value of metaphors with a view to simplifying the processes described (see section 6.2.2), and we can also point to its impact on the interpretation of analogical contexts. First, let us return to the analogy of conflict.

7.2.2.1. *Regulation instead of destruction*

The conceptual metaphor of conflict is prevalent in both corpora. We can even attest to the recurrent process of the extended metaphor, notably in certain passages with a particularly dense lexical field:

> 155) "Tremblez cochenilles et mineuses de tout poil, la *guerre biologique* vous est *déclarée*! Vendredi, un premier lâcher de larves d'*Exochomus quadripustulatus* (une coccinelle en français) a lancé les *hostilités* sur le parvis de la poste (place *De-Gaulle*) où deux tilleuls *moribonds* sont aux prises avec la *redoutable* cochenille pulvinaire. Rien de spectaculaire mais plutôt l'occasion pour le "*general*" Vernier

de passer ses *troupes* en revue et de rappeler tout le bien qu'il pense de *l'opération*"[42] (*La Voix du Nord*, 2013).

156) "*Warfare* on a college lawn. Ridding grass of crane fly larvae requires biological *weapons of mass destruction* [...] a *counter-weapon,* Sportnem H. Its packets teem with "entomopathogenic nematodes", ready for *attack.* At that season its agents are wondrous *warriors*" (*The Financial Times*, 2017).

While the use of a metaphor is relevant for simplification purposes, this rhetorical strategy has weaknesses in terms of interpretation. In this example, it is based on the fact that, in a conflict, two sides are opposed and only one of them will win. This principle is clearly expressed in the following extract:

157) "Quelles bêtes utilisez-vous pour *exterminer* les pucerons?"[43] (*La Nouvelle République du Centre*, 2011)

This assimilation is also found in English, with the very evocative analogy of war and that of the total destruction of "enemies" (*massacre, carnage*):

158) "A *KILLER* ladybird from Asia which *massacres* our native insects is invading British homes in its BILLIONS"[44] (*The Sun*, 2010).

159) "Now is the time to apply it to the soil and halt the *carnage*" (*The Guardian*, 2014).

However, as noted earlier (see sections 4.1 and 7.2.2), integrated pest management (IPM) methods are not intended to completely eliminate the presence of a pest in an ecosystem. The aim is to regulate pest numbers, a principle that is much more measured and complex than outright destruction. The term *régulation biologique* (*biological regulation*) would more accurately convey the original principle of IPM, that of limiting (not eradicating) pest populations. It should be noted that our corpus does not contain any occurrences of the term *régulation*

42 English translation: "Tremble, you mealy bugs and leaf miners of all kinds, *biological warfare* has been *declared*! On Friday, a first release of the *Exochomus quadripustulatus* larvae (*ladybug* in English) launched an attack on the square in front of the post office (place *De-Gaulle*), where two moribund trees are battling with the *dreaded* mealybug. It's nothing spectacular, but rather an opportunity for '*General*' Vernier to review his troops and to reiterate all the good he thinks of the *operation*" (*La Voix du Nord*, 2013).

43 English translation: "What critters do you use to *exterminate* aphids?" (*La Nouvelle République du Centre*, 2011)

44 The terms were in capital letters in the corpus.

biologique (or *biological regulation*), which is attested in LSP. This is the main entry proposed by the *Grand Dictionnaire Terminologique* to refer to the field studied. However, press discourse relies more on the better-identified term *lutte biologique* (*biological control*), leading to this pervasive extended metaphor.

This model, in which each combatant must defeat the other, actually corresponds to the principle of chemical plant protection methods, which has made it very successful in industrial agriculture: the fact that pesticides are wide-range, in other words, that there is not a specific selection of targets, and that the aim is total *eradication*. The issue is that chemical plant protection methods, which is a priori the model to be avoided, improperly becomes a discursive strategy applied to biological control methods, with a view to simplification (see Boutet and Parmentier-Cajaiba 2020). It is therefore not easy for the press reader to appropriate these complex notions.

We now turn to the analysis of another analogy, resulting from the binarism observed in the corpus.

7.2.2.2. *Good versus evil*

Some extracts contrast good insects with bad insects:

> 160) "Le biocontrôle, comme le *recours aux "bons" insectes*, aux champignons qui combattent les maladies ou encore aux huiles essentielles [...]"[45] (*La Nouvelle République du Centre*, 2017).

> 161) "Finally, to tackle the *dreaded mealy bug*, the *cute ladybird* species, *Cryptolaemus montrouzieri*, can be bought in little boxes of 25 individuals" (*The Guardian*, 2018).

While being explicit in French (which mentions "*bons insects*" or "good insects" in English), the second example in English uses a "cute ladybug" (*coccinelle coquette*) to defeat the "dreaded mealy bug" (*coccinelle redoutée*). From a scientific or biological point of view, there is no justification in opposing insects on the basis of a beneficial or harmful effect, as each insect plays a specific role in the natural ecosystem, according to its natural predation mechanism. Assessing identical objects (insect vs. insect) as good or bad has the rhetorical advantage of simplifying explanations, but does not accurately reflect the process of the food chain, as every insect is potentially the predator (or prey) of another.

45 English translation: "Biocontrol, such as *the use of 'good' insects*, disease-fighting champions, or even essential oils [...]" (*La Nouvelle République du Centre*, 2017).

This Manichean discursive situation can be expanded to an opposition between good and evil, which is also present in other situations. An illustration is presented here using two extracts from the corpus, in which the biblical reference to Armageddon – opposing the forces of good and evil – is used as the main term of a phrase, which is modified in turn by two opposing qualifiers (*ecological/chemical*):

> 162) "Last year scientists warned of "*ecological Armageddon*" as the number of flying insects has plummeted by 75 per cent in the past 27 years" (*The Independent*, 2018).

> 163) "[...] 1 per cent that survive the *chemical Armageddon*" (*The Daily Mail*, 2010).

This universal reference to Armageddon presented from two contrasting perspectives confirms the importance of binary and analogical reasoning in the process of explaining complex notions.

The study of the concepts of biological *versus* chemical pesticides, from the perspective of a conceptual metaphor (see Lakoff 1980; Kovecses 2005), deserves an analysis of its own, and is currently the subject of extensive study. We refer the interested reader to other ongoing research (see Monachon forthcoming).

7.2.2.3. *Summary and perspectives*

A summary of the issues observed emphasizes the following elements:

– The multiplication and simplification of designations by the press (based on lexical fields, synonymy, etc.) create an interpretative difficulty, which is related to the articulation between LSP and LGP.

– Associating explanations to socially shared cultural elements, spatio-temporal situations or binary framing (such as "pesticides are harmful", "biopesticides are virtuous") does simplify explanations, but also creates a potential difficulty in interpretation.

We have shown the difficult task of mediating a subject that is particularly complex from a terminological and cognitive point of view. As Morin (1990, p. 9) asserted, "science is intrinsically, historically, sociologically, and ethically complex"[46]. It is no longer a question of seeking to eliminate this complexity but to show it, by considering a series of perspectives in this domain, whether they are technological, environmental, bioethical, economic, legal or social. A simplistic vision of the issue of plant protection methods cannot therefore contribute to an

46 Original French version: "La science est complexe, intrinsèquement, historiquement, sociologiquement, éthiquement" (Morin 1990, p. 9).

efficient communication about this specialization. Without wishing to maintain a defeatist outlook regarding the possibilities of popularization or mediation of a specialized field (especially when the task assigned to science popularizers is too ambitious (Jacobi 2005)), we wish to further reflect on communication in this field. This is all the more necessary as biocontrol does not (yet) generate significant media coverage, and based on our current analyses, it also enjoys relatively "good press". However, we believe that this is just an emerging question, possibly behind the urgent coverage of other subjects (global warming, nuclear power, sanitary crisis, etc.). We cannot predict the future of biocontrol in the media; however, it is likely to become more important as society becomes more aware of global health issues (see *One Health*, footnote 11 in the Introduction), and the preservation of soils, rivers and crops.

In the last chapter, we wish to present our thoughts on such questions, as well as consider possibilities related to communication about these innovative plant protection methods and products, whose nature and properties are different from so-called "conventional" approaches.

8

Challenges in Agroecology

Complexity is beginning to appear not as the enemy to be defeated, but as the challenge to be met[1]

Edgar Morin (1990, p. 8)

As we have shown through our analyses, mediating agroecological issues is a heavy challenge. It is therefore necessary to think about innovative communication models outside the narrow circle of specialists.

First, it must be remembered that there is a shared responsibility when it comes to the mediation of scientific information, hence the importance of collaboration between the scientific world and the community of mediators.

8.1. A shared responsibility

A typical reader of the press does not read it to acquire scientific information, as was specified earlier (see section 1.2.1). In that respect, what is the journalist's mission, in relation to the reader, when dealing with agroecological issues: is it to inform, to enlighten, to seduce, to entertain or all of the above?

In order to better answer this question, let us observe the relation between the content and style of information within our double corpus.

1 Original French version: *La complexité commence à apparaître non comme l'ennemi à éliminer, mais comme le défi à relever.*

8.1.1. *The connection between content and style*

We illustrate our point with two themes that have been identified during corpus analysis: that of risk, which is related to a sort of sensationalism, and that of simplicity, and even an oversimplification of notions. Sections 8.1.1.1 and 8.1.1.2 outline and discuss these issues, which concern both the content of information and the way it is distributed.

8.1.1.1. *Risk and sensationalism*

We have observed that biological control tends to be presented in a positive light, if we consider its objectives or its "intentional effects": alternative methods in plant protection are developed with an initial virtuous intention, which is to replace chemically synthesized pesticides. However, these innovative methods also raise questions, related, for example, to the notion of risk in the manipulation of living organisms. The press thus refers to the processes involved in biocontrol, which consist of introducing an insect (potentially non-endemic)[2] into a geographical area to control a pest; another biocontrol method is the application of products (based on micro-organisms, pheromones or natural substances) to protect plants against pests. Although the principle of the developed methods is to reproduce plant protection mechanisms as they exist in nature, they involve human intervention in natural ecosystems – and therefore a potential disruption of these systems – which are likely to result in a range of "unintended effects" (ENI)[3].

This phenomenon can be illustrated by the introduction of the Harlequin ladybird (*Harmonia axyridis*) in Europe for biological control purposes, an insect whose development became a problem at the beginning of the 2000s[4]. Other types of ecosystem disruption exist (through the globalization of trade or unexpected movements of insects within containers, such as the Asian or Oriental hornet), but it seems less acceptable when there is a deliberate introduction of a species, as in this case here. Here are two extracts from the corpus, highlighting the issues presented by the Harlequin ladybird on European territory:

---

2 We emphasize that some forms of biological control, such as conservation biological control, aim to promote the development (or even the establishment) of "natural" and endemic enemies (that are already present on the territory in question).

3 ENI stands for "effets non intentionnels" in French. Various actors working on these processes are currently addressing this issue (see the ENI-BC network, which brings together researchers from INRAE, CIRAD and IRD, as well as various research laboratories studying all of these effects from a multidisciplinary perspective).

4 See: https://www.rhs.org.uk/biodiversity/harlequin-ladybird [Accessed January 17, 2024].

164) "Described as a '*voracious invader*' with a *frightening appetite* for other ladybirds [...], the harlequin causes *understandable alarm*" (*The Guardian*, 2016).

165) "*THE LADYKILLERS*; Invasion of Asian ladybirds that feast on UK bugs. A KILLER ladybird from Asia which massacres our native insects *is invading British homes* in its BILLIONS. The Harlequin ladybird – which also *bites humans*, emits a rotting smell and stains surfaces with orange colored goo – was one of the few species to thrive over the harsh winter" (*The Sun*, 2010).

The first extract describes an "alarming" invasion phenomenon, expressed by a specific lexicon (*voracious invader*, *frightening*, *alarm*). The second one, which addresses this same phenomenon, can be likened to a scenario from a dystopian movie[5], making the reader aware that the consequences of this invasion can affect them personally (*invading homes*, *biting humans*, etc.).

In these two examples, the mediator expresses the risk linked to "scientific progress", which is a sensitive subject to the general public. Certain studies even attest to a form of discomfort particularly in the controversial field of biotechnologies (see Frewer et al. 1997; Bauer 1998; Holmgreen 2008). The average citizen would thus be particularly concerned about scientific research using or involving living beings (especially animals) (Frewer et al. 1997, p. 100). This is the case with biological control, although biocontrol also include nonliving inputs (from living organisms or minerals). In any case, the public assessment of biotechnologies in this context seems to largely depend on "the relation between risk, benefit, and ethics" (Holmgreen 2008, p. 100).

This ethical issue can be brought closer to a more general debate related to human interference in natural phenomena. In this context, Moirand (2003) refers to the notion of interdiscursive memory, or collective memory, which no longer necessarily associates the subject with scientific questions, but rather with ideological, philosophical or ethical issues:

> This memory bank does not relate to an identifiable discourse source, nor to the sayings/words of particular speech communities, but instead to the ideological, philosophical and ethical positions of the discursively antagonistic groups: those who support science as an inescapable factor of progress versus those who support nature, for whom science is dangerous and destabilizing (Moirand 2003, p. 196).

5 The title of the article in capital letters is a real American movie title, released in 2004.

Opposition to science is the source of an "anti-progress" discourse, stigmatizing certain fields of research that have been deemed to carry risks. This verdict seems paradoxical, given that the explicit objective of these fields is to propose solutions that are less "harmful" to the environment, by attempting to reproduce natural mechanisms. Parrenin and Vargas (2020) observe this same paradox in the discourse on wind turbines. Scientific research aims to harness a natural phenomenon (the force of wind) to provide a sustainable energy source.

However, debates and conflicts are emerging on what can be considered as unintended effects that are potentially induced by scientific innovation. In the case of wind turbines, this can, for example, be the pollution of landscapes, noise pollution, the danger they represent for birds or the impact of the waves produced[6]. In the case of biological control, if we consider that the various unintended effects are part of what must be communicated, the way in which they are referred to will influence the layman's perception of the field. Therefore, the "communication risk" here would be to convey the potentially out-of-control invasion of non-endemic and threatening insects, resulting from the manipulation of living organisms through experimental research. In the face of such a representation, biological control could then be described as a "time bomb", and its researchers as "mad scientists", who are meddling with natural phenomena in a reckless and dangerous, even unconscious, manner. The "social meaning" attributed to these practices by the media, especially if they appear to be inadequately monitored, would then lead to other controversial cases involving principles that the average citizen is sensitive to (the precautionary and transparency principles)[7].

---

6 These criticisms, whether about wind turbines or natural plant protection methods, sometimes originate from discourses involving opposing technologies, or even from lobbies seeking to challenge the positive effects of innovation.

7 It should be noted here that some biocontrol products, which are considered as plant protection products (see Article L253-6 of the CRPM), are subject to general regulations and "are assessed in terms of their efficacy, selectivity and risks for humans (operator, worker, residues, etc.), the environment and non-targeted organisms that are present (water, air, soil, fauna, flora)". These products must therefore acquire marketing authorization (MA) to be sold. In addition, part of the research in this field aims to understand the dynamics of the populations, in order to understand the mechanisms of dissemination and thus the potential risks (which is not necessarily the case for the unintentional dissemination of insects through the transport of goods). However, we would like to point out that, as far as marketing authorizations are concerned, they do not involve macro-organisms, but only what the French Rural Code calls "*produits phytopharmaceutiques*" ("phytopharmaceutical products" or biocontrol products).

This notion of a "mad scientist" – and, more generally, the lexical field of madness – had already been observed in media discourse in a study of genetically modified organisms. Moirand noted an "imbrication between specialized terms (genetically modified soy) and mediated formulations (crazy soy)"[8] (Moirand 2007, p. 29). This lexical field arose from the emergence of Bovine Spongiform Encephalopathy (BSE), which was mediated under the name of "mad cow disease". The theme of madness was then transposed to other cases (chicken contaminated with dioxin became "mad chicken", polluted salmon became "mad salmon"). The exploitation of such a theme then became a mediation scheme, or even an argumentative strategy, which relies on the semantic play on words, especially in the case of emerging notions (Moirand 2007, p. 39). Other semantically related qualifiers are thus present in media discourse, such as *unhealthy* (*malsain*), *contaminated* (*contaminé*), or *uncontrollable* (*incontrôlable*), and stereotyped representations such as *playing with fire* or *manipulation* (in the figurative sense of the term) (Moirand 2007, p. 38). The potential problem of such communication strategies lies in seeing a particular theme take shape in qualifying (or non-qualifying) designations, through chains of intratextual and even intertextual references (Moirand 2007, p. 30). However, such a strategy also engages in a form of "seduction" in press articles, in the necessary appeal to potential readers, and thus in the necessary association between information and captivation (Charaudeau 2011).

This question of "seduction, specific to popular science discourse, has been studied for a long time" (e.g. in Winsten 1985; Nelkin 1986; Devries 1988; Charaudeau 2011; Bednarek and Caple 2014; Labasse 2012; Molek-Kozakowska 2017). Nelkin already alluded to this when discussing innovations in science: using "revolutionary breakthroughs" would be more likely to attract a reader than mentioning "recent findings" (Nelkin 1986, p. 113). On the contrary, as soon as it touches on a form of dramatization or sensationalism, the strategy is often criticized. Charaudeau summarizes the contradiction that is inherent in media information, based on two objectives, that of making people know ("*faire savoir*", in French), and making them feel ("*faire ressentir*"): the first one "should convey neutral and therefore credible information"[9]; the second one "should rely on appropriate

8 Original French version: "*Imbrication entre termes spécialisés (soja génétiquement modifié) et formulations médiatisées (soja fou)*" (Moirand 2007, p. 29). With release titles, such as *Alerte au soja fou*, as a play on words with the 1990's *Alerte à Malibu* series.

9 Original French version: "*faire savoir* [...] *qui devrait tendre vers un degré zéro de mise en spectacle de l'information et satisfaire au principe de sérieux en produisant des effets de crédibilité*" (Charadeau 2011, p. 74).

strategic choices of staging information and emotionally move the reader, by producing effects of dramatization"[10] (Charaudeau 2011, p. 74).

This contradiction is inherent to the "the commercial logic of the media" (see Sicard 2011); and also (Flaysakier 1997; Cabedoche 2003) with the aim of attracting as many readers as possible. According to Charaudeau, any media that "sticks to sober and rigorous communication is doomed to disappear"[11] (Charaudeau 2011, p. 75). Based on this principle, the author concludes that "media is thus bound to trigger readers' emotions, to mobilize their affect and spark interest and passion in the information that is transmitted [...]"[12] (Charaudeau 2011, p. 74). Labasse (2012) also defends this principle when the objective is to make a scientific fact *known*. Through his comparison between the film *Jurassic Park* and Stephen Hawking's book *A Brief History of Time*, he shows that, as far as science communication is concerned, the former has more of an impact than the latter. Although the film is considered as a form of entertainment (which casts a certain stigma on scientific research, leading to risky consequences), it does at least arouse the spectator's interest (and therefore curiosity) in a scientific subject (which would potentially encourage them to read further or obtain more information on the subject). Stephen Hawking's book, although of great scientific quality and written in layman's terms, has the weakness of only being addressed to a very limited readership. Various researchers collectively agree that the way in which information is reported in the press imposes formal aspects that are certainly attractive. Ruellan asserted the fact that journalists "consider themselves entitled to use – in moderate doses – the literary methods of reconstruction, narrative, representation, incarnation, dramatization, in order to make their effort to understand and restore reality more effective"[13] (Ruellan 1992, p. 35). Indeed, the various strategies mentioned can be considered as useful methods of "seduction" in terms of journalistic work, with the objective of producing "good stories that matter" (Secko et al. 2013, p. 69).

10 Original French version: "*faire ressentir* [...] *qui devrait tendre vers des choix stratégiques appropriés de mise en scène de l'information et satisfaire au principe d'émotion en produisant des effets de dramatisation*" (Charadeau 2011, p. 74).

11 Original French version: "*Ne se satisferait qu'à la rigueur sobre et ascétique du faire savoir serait condamné à disparaître*" (Charaudeau 2011, p. 75).

12 Original French version: "*L'instance médiatique est donc 'condamnée' à faire ressentir des émotions à son public, à mobiliser son affect afin de déclencher chez lui intérêt et passion pour l'information qui lui est transmise* [...]" (Charaudeau 2011, p. 74).

13 Original French version: "*S'estiment en droit d'utiliser – à dose modérée – les procédés littéraires de la reconstitution, du récit, de la fresque, de l'incarnation, de la dramatisation, afin de rendre plus efficace leur effort de compréhension et de restitution du reel*" (Ruellan 1992, p. 35).

We now move on to the second theme related to this field, affecting both the content and style of information: simplicity.

8.1.1.2. *Simplicity and lightness*

In Chapter 6, we showed discourse strategies based on analogical or binary reasoning that make it possible to explain a complex situation in simple terms (see sections 6.1.2 and 6.2.2). Other forms of writing in media discourse also emphasize simplicity (both of processes and reasoning), as shown in the following two extracts:

> 166) "*Exit donc pesticides* et autres désherbants totaux. Et *vive les bestioles*"[14] (*L'Est Républicain*, 2010).

> 167) "Got an alien invader plant species? Unleash an alien insect invader to fight back. One invader eats the other, and then abracadabra: problem solved. *That was the theory, anyway*" (*The New York Times*, 2011).

Any reader can be stunned and potentially seduced by such passages (or scenarios). These situations seem relatively trivial and allow access to simplified referents, as readers may expect :

> [...] the flippant and non-serious tone adopted in some of the articles in the science section of a daily newspaper such as *Le Monde* (absence of structuring; narrativization; use of general or trivial vocabulary, of verbs introducing quotations of a very subjective non-scientific nature: to deplore, to marvel at, etc.) can only be explained by the concern to attract readers (Beacco et al. 2002).

However, the situations described do not refer to processes as basic as they may seem. Thus, another "communication risk" involves giving the impression that the field presents a sort of facility, lightness, or even frivolity.

Following this line of reasoning (on the simple processes involved and the associated risk), some mediators remind us that humans have a duty to react (and "repair the damage") to a problem that they have caused through altering the global ecosystem between species:

> 168) "Species have always shuffled about to a degree. Yet many ecologists think that *our species has so altered the global biota* by

14 English translation: "*No more pesticides* and other total weed killers. *Long live bugs*" (*L'Est Républicain*, 2010).

> shunting insects, plants and microbes faster and farther than they could have moved alone that we have a duty, as Ricciardi puts it, "*to clean up our own mess*" (*The Financial Times*, 2016).

By using the common expression "to clean up our own mess", the journalist implies that it would be quite easy for humans to step backwards and avoid such a chaotic situation (or a "mess") which they are the cause of (in this case, the deliberate and problematic introduction of insects in a territory).

The two themes that are studied here (risk and simplicity) concern the unintended effects of scientific research in this field and the disruption of the natural ecosystem. To convey these factors, the journalist is caught between two potential approaches: the explanation of the complex processes involved in innovation (thematic framing), and a textual form that must be accessible and attractive (episodic framing). Two perspectives are being mixed here, the content being a priori guaranteed by the scientific world and the form being assured by the journalist. The role of each one is evidently a determining factor in this process of mediation, as the responsibility of the quality of information is shared.

Beyond the necessary collaboration between the scientific world and that of journalists, a second observation was made, confirming the shared responsibilities in the development of the domain under study.

8.1.2. *Development is inseparable from dissemination*

The second observation comes from teams of researchers involved in biocontrol, from very diverse backgrounds and disciplines[15]. Here is the context of these analyses, which also draw on the necessary collaboration between the scientific world and the world of mediators.

It is now well known that the use of phytosanitary products (also called pesticides) has an impact on the environment and, consequently, on human health. It is therefore important to consider a transition between the agrochemical paradigm and that of agroecology, by proposing alternative solutions to these controversial products. Research and development on natural protection mechanisms progress, with interesting and even increasing efficiency rates, especially in certain areas and under certain conditions. However, these innovative methods and products are struggling to be disseminated on a large scale. The Valbio and ENI-BC research communities have been addressing these issues since 2017, so as to identify the set

15 We refer to the Valbio and ENI-BC research communities.

of obstacles in the development of this field of expertise, according to different perspectives from the humanities and social sciences.

Hypotheses and experiments have already identified potential explanations for this problem of "socio-dissemination" of innovation (in sociology, economics, management, applied linguistics, etc.). These must now be the subject of more specific investigations and actions must be taken. Faced with these obstacles, and aiming to find ways to improve the dissemination of information, scientific mediation work is, at this stage, crucial.

This operation should maintain the existing collaboration between researchers of different disciplines, and the other parties involved in this field. The general perspective of this project is to include this field's communication strategy in the larger picture of information on products and methods that are part of innovation. It is also a question of proposing solutions for the variety of terms used, on how to adapt communication to the various audiences or, more generally, on how to provide new communication tools for the popularization of this specialized field, in particular concerning its unintended effects[16].

This principle of collaboration between the scientific world and that of mediators partly follows the view supported by Peters (2012), who recalled the concept of the "medialisation" of science, coined by the German sociologist Weingart (2001). The objective is to promote a stronger relationship between the world of science and that of the media, as this principle of mutual reliance, which was already brought to light by Nelkin (1986, p. 8), should accompany a process of remediation in the face of the mutual distrust that may exist between these actors.

In this final section, we have grouped together a set of ideas related to the mediation of science in this field of specialization.

8.2. Scientific mediation

We are aware that the general daily press does not have the same objectives as the specialized popular science press. This book mainly focuses on the general press. However, in this third part, we chose to broaden our approach by considering the possible convergences between these two types of journalism (general and specialized) and propose different communication strategies, on both scientific and

16 These thought processes are being discussed and conducted by many multidisciplinary actors within the Valbio and ENI-BC collectives. One of the first missions of the ENI-BC collective is to simply establish the list of what can be characterized by the term "*effet non intentionnel*" ("unintended effect") of biocontrol and stabilize these aspects, so that effective communication can be carried out on these subjects.

environmental issues, to a non-expert audience. These strategies involve a more general line of questioning, which concerns the social function of scientific mediation (Lieutenant-Gosselin et al. 2012; Lipani and Pascal 2020; Naït-Bouda and Pelissier 2021) as well as the co-construction of the image of science by the mediator and the layman.

We have identified three themes that, to us, seem to be decisive in this context: the first involves the necessary reflexivity for the mediation of science; the second involves the language used for this purpose; the third widens the perspective to the issue of audience and their involvement in this process.

8.2.1. *Reflexivity and scientific mediation*

We have shown the recurrent theme of risk in this specialized field, as well as the delicate communication on this subject. This is all the more complex as the notion of risk is associated with a phenomenon that is inherent to scientific research: uncertainty.

8.2.1.1. *Scientific method and uncertainty*

Considering uncertainty to be a strength in communication (instead of a weakness) is a principle supported by many researchers (such as Friedman et al. 1999; Dunwoody 2004; Dunwoody et al. 2018; Kampourakis and McCain 2019). The aim is to provide an understanding of the concept of risk, through a deeper comprehension of the methods being developed. As Nelkin previously stated, "while we welcome technology as the key to progress and the solution to problems, we are increasingly preoccupied with risk, fearing the very technologies we depend on the most" (Nelkin 1986, p. 7). It is therefore a question of explaining the notions of doubt and potential risk, which are characteristic of the scientific approach. This principle is opposed to the strategies used in cases of scientific controversy, where the notion of doubt, or risk, is exploited as a weakness of scientific reasoning, calling attention to debatable or false arguments (Oreskes and Conway 2012). Indeed, science accepts doubt as long as a consensus is not established, but not as the support for asserted or opposing positions on subjects that are still under study.

In terms of social acceptability, it therefore seems important to us to focus part of biocontrol communication on some of its unintended effects, so as to anticipate fears, distrust or potential controversies. Raichvarg and Jacques have emphasized the fact that "a world [that] we understand better is a world we are less afraid of"[17] (Raichvarg and Jacques 1991, p. 30). With this objective in mind, and following

17 Original French version: "*Un monde où l'on comprend mieux est un monde dont on a moins peur*" (Raichvarg and Jacques 1991, p. 30).

various studies (e.g. Burns et al. 2003; Dunwoody et al. 2018; Klein 2020a), we believe that the public should not only have access to the objectives and expected results, but also to the methods implied by scientific research (see section 1.3.1). These methods are based on observation, experimentation, doubt and lengthy periods of time, which is what gives them their strength and legitimacy. For a researcher, "the incompleteness inherent in knowledge"[18] is natural (Laügt 2001), and inevitably leads to a form of uncertainty (particularly in the short-term) or even risk. By denying this risk, we are questioning the scientific process itself. If, on the contrary, the citizen understands the importance of time dedicated to research and experimentation, they should also be able to assess the "risks of progress", and even accept them.

This requires a form of reflexivity on the field itself, its objectives, its issues and its methods. This becomes more complex when we are faced with a particularly multidisciplinary situation. As we indicated earlier, plant protection by natural mechanisms involves many disciplines, and the terms used do not always cover the same elements according to the perspectives[19] (Bernard 2017). While it is not possible for a journalist of the daily press to delve into disciplinary discrepancies, we can nevertheless assume that the mediator can take on the privileged role of providing a kind of hindsight on multidisciplinary practices and issues. Jurdant underlines this characteristic of popular science discourse, as opposed to scientific discourse, often blamed for the absence of reflexivity (Jurdant 1973, p. 143). Jeanneret also believes that the benefit of science popularization is that it cannot quite "erase the various points of view, nor the fact that they have emerged in the course of history"[20] (Jeanneret 1994, p. 128).

8.2.1.2. *Specialized field and disciplines involved*

Agroecology is a scientific field involving many disciplines (biology, chemistry, agronomy, plant pathology, biochemistry, entomology, etc.). Environmental and scientific communication must also integrate numerous perspectives (social, economic, legal, political, ethical, etc.), which go far beyond the disciplinary field of exact sciences. It is a question of creating a dialogue between disciplines, with their collaboration then being considered as a strength. The need for interdisciplinary dialogue has long been established and the issues are ever more prevalent (e.g.

18 Original French version: "*L'incomplétude inhérente au savoir*" (Laügt 2001).

19 Original French version: "[…] *les termes employés ne recouvrant d'ailleurs pas toujours les mêmes éléments en fonction des perspectives*" (Bernard 2017).

20 Original French version: "[…] *effacer la pluralité des points de vue, ni le fait qu'ils ont émergé au fil d'une histoire*" (Jeanneret 1994, p. 128).

Schoenfeld 1980[21]; Nelkin 1986; Burns et al. 2003; Galochet et al. 2008; Jollivet 2008; Priest 2010; Trench and Bucchi 2010; Bourassa et al. 2013; Libaert 2016; Chlous et al. 2017; Catellani et al. 2019). Jollivet describes the interest of confronting disciplinary points of view: "Each of them brings its own analytical strength, which enriches the analysis; but on the other hand, their confrontation forces one to go beyond the partial – and therefore biased – nature of the explanation that their presuppositions lead to"[22] (Jollivet 2008, p. 16). Chlous et al. (2017) confirm this principle of focusing on "a discussion between approaches that fall under different paradigms and theoretical framings, wagering that this confrontation enriches the discussions"[23].

This intention is, for example, revealed through national calls for research projects (such as ANR projects[24], that tend to favor interdisciplinary projects). Other initiatives (see Galochet et al. 2008; Chlous et al. 2017) make it possible for interdisciplinary collaborations or research laboratories to emerge, bringing together specialists from various fields around jointly defined focuses and objectives (e.g. the LIED and COSTECH[25] research centers that respectively work on energy transition and technological innovation in a resolutely interdisciplinary framework, as quoted by Pecman (2018, p. 36)). The Communication, Environment, Science and Society research group[26], which we are a part of, is also bearing witness to this multiplicity of disciplines (information and communication technologies, linguistics, sociology, etc.), converging for a better understanding of environmental issues.

More recent models of science communication (see section 1.2.2.3) therefore move away from traditional models, relying on a diversity of scientific knowledge sources, and not just on the so-called "hard" sciences (Secko et al. 2013). In these

---

21 The latter already declared the following in 1980: "Environment journalism is not a clear-cut beat but rather represents the 'business-medical-scientific-economic-political-social-pollution story'."

22 Original French version: "*Chacun d'eux apporte sa puissance analytique propre, ce qui enrichit l'analyse; mais par ailleurs, leur face-à-face oblige à un dépassement du caractère partiel – et donc partial – de l'explication à laquelle les conduisent leurs presupposes*" (Jollivet 2008, p. 16).

23 Original French version: "*Une discussion entre des approches qui relèvent de paradigmes et de cadrages théoriques différents, faisant le pari que cette confrontation enrichit les discussions*" (Chlous et al. 2017).

24 Source: https://anr.fr/en/.

25 The LIED is attached to the Université de Paris (see: http://ed382.ed.univ-paris-diderot.fr/LIED), and COSTECH to the Université de technologie de Compiègne (see: https://www.costech.utc.fr/Le-COSTECH) [Accessed January 17, 2024.].

26 See: https://comenvironnement.hypotheses.org/le-ger-communication-environnement-science-et-societe [Accessed January 17, 2024.

innovating models, “a distinct difference can be found between science being the legitimizing form of knowledge among traditional models, and science being seen as equal to other forms of knowledge in the case of non-traditional models”.

The various disciplinary contributions thus make it possible to broaden the horizon of research, by adopting a more systemic position (Morin 2005; Jollivet 2008), bringing new thought processes to light (Libaert 2016; Catellani et al. 2019), and even a new unified discipline (Boutet and Parmentier-Cajaiba 2022).

Why then is it so complex, and why is working across disciplines still a challenge?

Bauer et al. partly answer this by mentioning the “lack of common foundations” for such practices (Bauer et al. 2007, p. 87). “[...] As science communication is inherently both multidisciplinary and interdisciplinary, organizing the basic requisite knowledge at the core of its study is a particular challenge. Indeed, there is not a single ‘canon’ of literature or research that is prescribed for aspiring science communication students” (Priest 2010). There are many research and teaching situations that show that there is a difficulty in getting different disciplines to collaborate, especially from a methodological point of view (see, for instance, the assessment of research by peers from the same discipline[27], the lack of a disciplinary anchor for fields such as tourism or – more precisely – environmental issues). The history of science has seen this evolution: we have gone from the humanistic scientist who straddled several sciences (such as Leonardo da Vinci) to an expert in a single field of specialization. Indeed, the hyper-specialization in all fields of knowledge today does not necessarily allow scientists to position themselves in relation to other domains in science (Thouin and Lapointe 2016)[28]. According to these authors, “it is important that scientists also acquire a broad general culture and a broad scientific culture, the latter being in fields outside their area of expertise, which would allow them to more easily locate their research within a general context”[29].

27 From a pragmatic point of view, the assessment of research is not done on the basis of empiricism or interdisciplinarity, but often remains monodisciplinary.

28 Thouin and Lapointe (2016) illustrate this hurdle: “For example, even though the scientist may have lost sight of it a bit, the study of neuron clusters that are the cerebral ganglia of the sea slug is paving the way to a better understanding of how memory and the brain work”.

29 Original French version: “*Il importe que les scientifiques acquièrent eux aussi une vaste culture générale et une vaste culture scientifique, cette dernière dans des domaines autres que leur domaine de spécialisation, ce qui leur permettrait de pouvoir situer plus facilement leurs recherches dans un contexte general*” (Thouin and Lapointe 2016).

Another element of reflection is the difference between multidisciplinarity and interdisciplinarity. While the plurality of disciplines is implicit behind each term, their difference is explained as follows:

> *Interdisciplinary* projects, programs, and fields are generally defined as those that bridge, integrate, or combine work from more than one traditionally defined discipline, whereas the term "multidisciplinary" more commonly is restricted to the description of activities that require input from different kinds of experts but in more of a collaborative, rather than an integrative, sense (Priest 2010).

The same author illustrates this principle of knowledge integration through the case of a scientific committee, charged with studying the risk of a medical treatment on a child's pathology:

> To determine the cause of the disease could require an inter-disciplinary approach, one that not only takes various forms of expertise into account but develops a research plan that integrates some of these fields more closely (for example, by considering both genetic and nutritional or other environmental factors as variables in a single study). Science communication encompasses variations of all of these, united more by its subject matter than its methodology.

This indispensable networking of researchers should ideally be supported by the development of interdisciplinarity within publications (Galochet et al. 2008, p. 274), as is the case with a growing number of publishers who encourage these connections between disciplines, as well as journals that provide a forum for interdisciplinarity (such as the *Natures Sciences Sociétés* journal, the *Vertigo* journal, as well as the *Développement durable et territoires* journal) (Galochet et al. 2008, p. 279).

The sociologist Nelkin compared interdisciplinarity to a "team sport"[30]. This necessarily implies close cooperation and mutual respect between actors with very different backgrounds and profiles. In this team sport, the collaboration between journalists and scientists is not only crucial for the necessary reflexivity in this complex field, but also for better communication outside the experts' field. Within the framework of this collaboration, the language used remains central, taking into

30 Quoted by Priest (2010, p. 3): "[…] As the late sociologist of science communication Dorothy Nelkin once noted (in conversation), interdisciplinarity is rarely accomplished by a single individual's mastery of more than one discipline – by someone's becoming an 'interdisc', as she jokingly put it, although this certainly does happen on occasion. Rather, interdisciplinary inquiry is more commonly a sort of team sport – a matter requiring close cooperation and mutual respect among a group of people with diverse training".

account the strong interrelation between the content and the form of the information being disseminated. Let us now address the importance of the language and terminology used for communicating with or for non-experts.

8.2.2. *Language and terminology*

Media discourse analysis has brought to light the instability in the interpretation of certain specialized terms via press discourse. The specialized meaning of a term is not necessarily the same when it circulates outside the scientific domain. It is, moreover, influenced by the reader's interpretation. Indeed, as we have indicated in section 7.1.2, the "common" meaning of a lexical unit could already be well established, and its coexistence with the specialized meaning could then potentially lead to confusion for a non-specialist reader. This is the case for the term *lutte* (*control*), which, in French, evokes the notion of fighting, whereas the processes involved in *lutte biologique* (*biological control*) refer to a more technical sense of regulation. Another example is the term *pesticide*, which unfortunately carries a negative connotation because it is equivalent to *produit phytosanitaire* (*phytosanitary product*), and yet potentially refers to a natural product. Once the collective imagination has built its perception on a pejorative meaning, it can be difficult to "reframe" readers by explaining that biological control is a process of regulating numbers, and that pesticides, whether they are biological or chemical, should not be linked to any connotation from the outset (positive or negative). Faced with these difficulties, Meilliez reminds us of the importance of agreeing on the meaning of words and phrases, as "an indispensable prerequisite for any intellectual construction". What could be taken for granted is in fact a challenge, as it "requires an unshakeable will, a sense of listening that is constantly on alert, and constant vigilance"[31] (Meilliez 2008, p. 173).

This necessary vigilance implies holding and maintaining regular and frequent exchanges between the parties involved. However, this awareness, which involves close collaboration, is not always a realistic hypothesis, as the processes involved are constantly evolving and the exchanges are time-consuming.

Thus, the existence of an external resource could constitute a form of collaboration between scientists and "communicators", as a way of accompanying the journalistic work, or even the interpretative work of a heterogeneous readership. This would involve establishing a digital communication strategy and the creation of a visible, reliable and regularly updated support (Delforce and Estienne 2011),

31 Original French version: "[…] *un préalable indispensable à toute construction intellectuelle.* […] *nécessite une volonté inébranlable, un sens de l'écoute sans cesse en éveil, une vigilance permanente*" (Meilliez 2008, p. 173).

which would be accessible to the specialist, the semi-specialist, as well as the non-specialist.

A key element at the heart of these resources is the terminological and cognitive issues related to this area of expertise.

8.2.2.1. *Terminology and definition resource*

One of the primary objectives of this resource would be to (re)establish the terminology and definitions in this field, with a view to coordinating communication through different devices and supports.

The purpose of such a resource would be to:

– list the important terms of the field in the different languages of international scientific communication;

– define the terms in different languages and propose explanatory diagrams on what they cover;

– propose bilingual (or even multilingual) glossaries;

– explain the terms in their technical meaning. For example, specify the differences between a pesticide, a biopesticide and a biocide; between an insecticide and a natural or biological insecticide; or the characteristics of a chemical or biological product or a biochemical mechanism.

For example, INRAE press releases present interesting sources for terminology (as well as scientific method explanations). If we take the example of a press release on biocontrol methods[32], we find very precise, technical and professional terminology (e.g. *autoclave, zone confinée, personnel habilité, sas d'entrée*, etc.[33]), which would benefit from being quoted, defined and even explained and made accessible for various audiences.

An article on biological control could then build on these external elements, either by integrating them into the article in a specific box or by referring to existing resources. This cognitive contribution in particular formal aspects (boxes proposing names, definitions, mini-glossaries or reference to alternative sources) could be the deciding factor in reducing cognitive difficulties in the circulation of specialized information. This is part of a didactic approach, which is important for the dissemination of all scientific information.

32 See: https://www.inrae.fr/actualites/entretien-marcel-thaon-eleveur-dinsectes-inrae [Accessed January 17, 2024].

33 English translation: *autoclave, confined area, authorized personnel, airlock*, etc.

The same objectives can be found on the issue of definitions. Various studies have shown the absence (or lack) of official definitions in press articles dealing with scientific topics, such as Garric (2018) for earthquakes, Beacco et al. (2002) for bovine spongiform encephalopathy and Ledouble and Cadel (2022) for biocontrol. While this absence can be explained "insofar as the author assumes a shared stereotypical knowledge with their readership" (Beacco et al. 2002, p. 294), a reference to one (or more) "official" definition(s) would be useful in little-known fields, so as to avoid three potentially problematic situations:

– the absence of definition (Beacco et al. 2002; Garric 2018);

– the multiplication of definitions (Ledouble 2021; Ledouble and Cadel 2022);

– the competition between definitions ("definition struggles"), which are often at the root of the emergence of "public problems" (Gilbert and Henry 2012).

Given that biological control has been particularly affected by latent instability for several decades, a terminological and cognitive clarification seems crucial.

In addition to definitions, resources could be enriched using other ways to explain these topics. Several research seminars on these issues suggested different paraphrases to explain biocontrol to a wide audience, which would contribute to this objective of simplification:

a) Agricultural protection solutions aiming at a low impact on living beings.

b) A set of alternative methods to synthetic pesticides used for the prevention of disruptions linked to agricultural, livestock and environmental parasites. A neologism has been suggested in this context, that of "bio-prevention".

c) A set of instruments/methods/tools/solutions/strategies/techniques for the control/management/regulation of species that are considered to be harmful, with a view to protecting plants.

These simplification efforts, resulting from interdisciplinary work between agronomists and researchers in humanities and social sciences, need to be developed. The objective is to establish simple and clear communication. In addition to the terms used in this mediation process, language structures also need to be looked at.

8.2.2.2. *Plain language*

The emerging notion of "plain language" (or "clear communication") seems to be an interesting avenue for us to explore. The objective of this language is defined as follows by the PLAIN network, which is in charge of its development and promotion: "Communication is clear when it is written, structured and presented in such a way as to allow the reader to easily find what he is looking for, to understand what he finds and to make good use of the information"[34]. The objective of this language, which is in constant development, is primarily to propose simplified syntactic and stylistic elements, but it also applies to the lexical and terminological choices used. Its application to the field of biological control deserves particular attention. We illustrate the principles of plain language below, in other contexts such as the medical, administrative and legal fields.

The Cochrane project proposes Plain Language Summaries (PLS)[35] to help non-specialists understand scientific articles without academic training. There are guidelines for setting up this type of communication, but in linguistic terms, the textual genre of PLS is poorly characterized. However, Gledhill et al. (2019) have revealed key simplification strategies used by medical experts when writing for the general public. Examples include using an interrogative structure in titles, instead of the *Main results* or *Authors' conclusions*, as illustrated in Table 8.1.

Abstract	PLS
Main results	***What*** *did the studies show?* ***How*** *well does it work?*
Author's conclusions	***What*** *is the bottom line?*

Table 8.1. *Use of how and what in PLS titles*[36]

It can also involve the reader by using the *you* or *your* pronoun (see Table 8.2), or rephrasing the statement by mentioning people or citizens involved in this study (see Table 8.3).

Abstract	PLS
Osteoporosis is a condition resulting in an increased risk of skeletal fractures due to a reduction in the density of bone tissue.	*Bone is a living, growing part, of* ***your*** *body. Throughout* ***your*** *lifetime, new bone cells grow [...] When* ***you*** *have osteoporosis, the old bone, breaks down faster [...].*

Table 8.2. *Use of the you/your pronoun*

34 See: https://plainlanguagenetwork.org/plain-language/what-is-plain-language/ [Accessed January 17, 2024].

35 See: https://community.cochrane.org/style-manual [Accessed January 17, 2024].

36 All tables have been taken from Gledhill et al. (2019).

Abstract	PLS
Age-related cataract is the opacification of the lens, which occurs as a result of the denaturation of lens proteins.	***As people get older**, the lens in the eye can become cloudy – this is known as a cataract.*

Table 8.3. *Mention of the persons or citizens concerned*

This can be achieved by systematically expanding the complex terms in order to explain a process (see Table 8.4) or by adding a relative clause with this same objective (see Table 8.5).

Abstract	PLS
Injection site reactions	*Skin reactions **at the site of** the injection*
Near the bladder neck and proximal urethra	***Around the junction between** the bladder and the urethra*

Table 8.4. *Development of complex terms from prepositional structures*

Abstract	PLS
Number needed to treat (NNT)	*The number of people **that need to be** treated*
Higher dietary levels of antioxidant vitamins	***People who eat** a diet rich in antioxidant vitamins*

Table 8.5. *Development of complex terms from relative structures*

Finally, the point of plain language is also to make the presentation of numbers less complex, which particularly applies to the results section (see Table 8.6).

Abstract	PLS
At least 50 % pain relief	***Good levels of** pain relief*
Congestive heart failure: RR 5.11	*Risk of heart toxicity is **five times more likely***

Table 8.6. *Lexicalized presentations of numerical comparisons*

The use of plain language is also studied in its application to the field of administration. The approach consists of making administrative procedures easier which, in many situations, call for a specific terminology. The objective is thus to reduce disparity among citizens in this context, as proposed, for example, by Bouchet (forthcoming) for the digital discourse of access to social rights, or Rossetti et al. (2020) in order to help the elderly understand financial information, or Clerc (2019) who describes the importance of the user's needs in a digital environment.

Fernbach (2010) illustrates the numerous initiatives to implement plain language in the legal field. A doctoral thesis (Bouyé 2022) looks at asylum procedures in English and French law. While these approaches facilitate access to information, other studies also have the goal of creating useful resources for users in complex fields. François et al. (2018), for example, propose a tool to assist administrative text editors in simplifying their writing: the AMesure platform[37].

The use of plain language has also become more common since the Covid-19 health crisis. Some institutions, such as the Swiss Federal Office of Public Health[38] and the Government of Canada[39], have proposed communication measures on their websites. Researchers working in interdisciplinarity also propose webinars on ethics in pandemic situations, using plain language[40].

This information can also be accessed by people with physical or cognitive disabilities, like the information published on the pandemic in the United States by the U.S. Government Office for People with Developmental Disabilities[41]. In addition, there are resources that answer frequently asked questions on the subject, published by the American Institute for Research (AIR), which is dedicated to improving the citizens' quality of life[42].

The same principle therefore underlies all of these observations, namely the necessary coordination between the work of researchers and that of communicators, for a qualitative approach to both the content and style of information. Though this collaboration "upstream" of dissemination is relevant, we also wish to achieve a

---

37 See: https://dial.uclouvain.be/pr/boreal/object/boreal:193864 [Accessed January 17, 2024].

38 See: https://www.bag.admin.ch/bag/fr/home/krankheiten/krankheiten-im-ueberblick/coronavirus/covid-19/barrierefreie-inhalte/leichte-sprache.html [Accessed January 17, 2024].

39 See: https://accessibilite.canada.ca/ressources/lignes-directrices-communication-accessible-organisations-urgences [Accessed January 17, 2024].

40 See: https://msss.gouv.qc.ca/professionnels/ethique/serie-de-webinaires-sur-l-ethique-de-la-recherche-en-contexte-de-pandemie/a-propos/ [Accessed February 2, 2022].

41 See: https://opwdd.ny.gov/coronavirus-guidance [Accessed January 17, 2024]

42 See: https://www.air.org/resource/brief/covid-19-frequently-asked-questions-plain-language [Accessed January 17, 2024].

form of collaboration "downstream", as well as take a look at the public's role in this co-construction of the meaning of disseminated information. We now address the question of the public's involvement in the interpretation of scientific information, as reported by the media.

8.2.3. *Involving the public*

The issue of public involvement opposes the initial deficit model of science communication, which viewed the public as lacking the knowledge to analyze and interpret (see section 1.2.2.3). As stated by Trench and Bucchi (2010),

> it is notable that in fields of communication closely related to science communication, such as risk communication, health communication and science education, similar discussions and reflections on models have taken place over a somewhat earlier period, though these developments have happened largely without reference to each other. In each case, the audiences or publics and their needs were increasingly foregrounded, as top-down models were progressively displaced or supplemented by more inclusive, egalitarian or participatory models.

For more than a decade, a consensus has thus been forming around the idea of collaborating with and even involving the public, and is currently at the heart of most studies on the renewal of generalist and scientific journalism (e.g. Brossard and Lewenstein 2010; Priest 2010; Bousquet et al. 2015; Secko et al. 2013; Bucchi 2016; Catellani et al. 2021).

Let us return to the importance and the positioning of the public in the process of science mediation.

8.2.3.1. *The public's positioning*

Regarding the coverage of an event by the regional press, Devries already wrote the following in 1988: "A local newspaper needs a local angle" (Devries 1988, p. 888). The principle at the heart of this approach is the readership's involvement through its proximity to the subjects being covered. This makes it possible to develop local interest in a theme that is potentially part of a more global societal context, as well as to rally a community of readers around it (Ringoot and Rochard 2005). It is by following this "think globally, act locally" tenet that the French regional press has long remained one of the preferred information channels in France (ACPM 2016).

While the regional press, in its traditional form, is sometimes perceived as the remnants of an old model (Ballarini 2008), it is diversifying its communication strategies in France, in the face of a general decrease in written press readership. The relationship with the public is thus based on multiple forms and innovative devices. New actors are continually emerging on the Internet, whether they are online pure-players who now share the monopoly of the regional press on the hyperlocal information sector (Bousquet et al. 2013), blogs or media that allow exchanges between the actors involved in the subjects being covered (such as *The Conversation France*), crowdsourcing initiatives, etc. Public involvement can also be achieved through the identification of local problems and the search for solutions, like in "solution journalism" (or other related terms such as "positive journalism" or "constructive journalism"). These forms of journalism aim to analyze practices or initiatives, which aim to provide concrete answers to societal problems[43]. They also make it possible to put the local area in the limelight by valuing the territory and its actors and to embody information by integrating the readers' daily lives (Amiel 2020). They also aim to stimulate their imagination, by reporting on both the complexity and resilience of the world (Vanderpooten 2020). In this context, there is an emphasis on interactivity. The journalist involves the reader and encourages his participation (through testimonies, contests, educational games, tests or even videos). The layman then becomes a producer of information and even of knowledge.

These strategies are, without question, of interest in science mediation. Thouin and Lapointe (2016) emphasize the fact that by "presenting applied research at the local scale, it is easier to stimulate the public's interest and to create an open discussion on more abstract and universal issues"[44]. Public involvement is recognized as a determining factor in understanding and interpreting certain scientific and social facts, as well as its participation in the experiments that facilitate the development of scientific knowledge. "Experiential knowledge" even becomes a prerequisite for acquiring certain knowledge, particularly in the humanities and social sciences (Gardien 2017). In this regard, the author mentions Dewey's position[45] of how scientific knowledge is second-hand to experiential knowledge, the latter being the primary point of interest in the development of scientific knowledge.

43 See: https://reportersdespoirs.org/sojo/definition/ [Accessed February 2, 2022].

44 Original French version: "*Présentant des recherches appliquées à l'échelle locale, il est plus facile d'éveiller l'intérêt du public et de susciter une ouverture à des problématiques plus abstraites et plus universelles*."

45 He has largely challenged these questions in his book *How We Think: A Restatement of the Relation of Reflective Thinking to the Educative Process* (first published in 1910).

It is therefore important to involve the citizen, both in the content and the style of information, by means of different mechanisms. For example, Secko tests different models of "experimental journalism" in order to observe their impact on readers. He proposes two "artificial" articles on the same subject (salmon genomics), according to two competing models: the scientific literacy model and the interactive science model; the first model relies on a vertical transmission of knowledge, aiming to bridge the gap between experts and laymen. The second model assumes "a more active didactic position, a framing of the issue as a social dilemma, and a more open stance on possible analyses and solutions"[46] (Secko 2012, p. 66).

There are also other practical experiences of involvement or collaboration with the public, such as that proposed by Valenti and Tavana (2005), which brings journalists and scientists together during a study in the field of tropical ecology. The journalists are involved in the scientific process, in connection with the surrounding nature and culture associated with local practices, (for instance, some knowledge of the fauna and flora at the origin of medical or scientific advances in the Western world). Witnessing the consequences and impact of actions on natural resources establishes the journalists as learners, enriching their own training from practical and personal observations and experiences. Babou (2017) confirms this critical perspective, believing that being an expert does not replace some form of investigation and experience by the public.

> It is this practical 'experimentation' of the experience of "others" and their environments, through the sharing of a job, a daily life, a leisure activity, or any other means to be imagined, that would then be the source of the decision, and not just the knowledge of experts [...][47].

The layman would then be confronted with the methods of observation and experimentation, within the framework of scientifically supervised activities.

Without delving into these issues here, we refer to the multidisciplinary work that highlights these principles, in the very broad framework that we can call "citizen science projects" (e.g. Catellani et al. 2021), and we also specify the need to develop them, particularly in the field of environmental communication (e.g. Couvet and Teyssèdre 2013; Chlous et al. 2017). The layman can thus take on different

---

46 Original French version: "*Une posture didactique plus active, un cadrage de l'enjeu comme un dilemme social et une position plus ouverte quant aux analyses et aux solutions possible*" (Secko 2012, p. 66).

47 Original French version: "*C'est cette expérience concrète de l'expérience des "autres" et de leurs milieux, passant par le partage d'un travail, d'une vie quotidienne, d'un loisir ou par tout autre dispositif à imaginer, qui serait alors la source de la décision, et non le seul savoir des savants* [...]."

postures, ranging from a simple data provider to an active contributor of the scientific project, a position "measured" by Millerand (2021) along an axis expressing a form of involvement. The author brings the first posture (data provider) closer to the deficit model, where the layman's participation in the project allows them, at the very least, to become aware of the scientific activity. At the other end of the scale, the layman holds his own knowledge, contributing to the success of the scientific project, and is not just a simple promoter conforming to the scientific protocols that are being implemented.

However, Priest acknowledges that "this is good advice but adds new layers of complexity to both the understanding and the practice of science communication". He mentions that

> few sound mechanisms exist within modern democracies for actively incorporating popular thinking into governmental and regulatory decision-making, especially for the most complex and novel policy issues presented by emerging scientific developments. Both academics and practitioners struggle to meet this added challenge (Priest 2010, p. 2).

There is another model of communication that places the layman at the heart of the transmission process, which does not require the intervention of a mediator. This is the alternative "deviation model", as advocated by Bucchi (1996) and Davies (2008), encouraging the scientist to speak directly to the target audience (see section 1.2.2.3).

This communication, in which the scientist is the mediator, is unevenly practiced across scientific fields. In the case of biocontrol, the dissemination of information still seems relatively limited. However, there are a few examples of these speeches online (such as TED talks)[48], or during conferences as part of Science Week, or even workshops that are devoted to introducing science to the general public. All of these initiatives need to be pursued.

However, as we have already pointed out, exchanges at any level of the information chain are not always easy. Thus, external resources – potentially multimedia and interactive – could also contribute to enriching these collaborative relationships, as well as present themes or problems specific to the field or even scenarios allowing for reflexivity by a variety of audiences.

---

48 For example, there is a mediation by Shimon Steinberg, a specialist in natural plant protection mechanisms. See: https://www.ted.com/talks/shimon_steinberg_natural_pest_control_using_bugs?language=en [Accessed January 21, 2022].

8.2.3.2. *(Inter)disciplinary perspectives and experiential scenarios*

The practice of reflexivity imposed by interdisciplinarity requires us to consider the existence of the disciplines at the origin of these discussions. As Jollivet reminds us, the aim is to look back at the scientific culture of the different disciplines that can be mobilized, as well as at their theoretical orientation, in order to study a question. Thus, the practice of self-reflexivity carried out by the specialists themselves must be subjected to questioning by researchers from the associated disciplines (Jollivet 2008, p. 15).

Without falling back into this monodisciplinary confinement that we have discussed earlier, one of the objectives of these external resources would be to recall the whole range of perspectives involved in this field and to express a variety of points of view. The following are a series of potential viewpoints to consider:

– environmental: impact of these methods and products on fauna and flora and on the environment in general;

– agronomic: knowledge of the methods, characterization of their effects;

– economic and financial: cost and yield of agronomic practices, linked to their efficiency;

– legal: the issue of manipulation of living organisms, associated patents and marketing authorizations for different substances and products, local and European regulatory levels;

– spatio-temporal: agricultural practices and historical evolution;

– sociological and societal: diffusion of new practices, social acceptability, impact on the life of citizens;

– philosophical and ethical: relations between human beings and nature, interference in natural phenomena;

– etc.

The aim is therefore to compare these points of view and to recall the significant complexity of both the reasoning and the processes involved within each discipline. The analysis can then draw on the theorized reflective approaches that are specific to different social sciences, with the approach then integrated into the research process itself (Chlous et al. 2017). This work is likely to highlight the fact that there can also be a form of heterogeneity inherent in each discipline. Jollivet reminds us that,

> far from having the epistemological unity that their title gives them, these disciplines are spaces of controversy and bring together extremely diverse research angles that may contradict with one

another and may even simply ignore one another[49] (Jollivet 2008, p. 13).

He illustrates this phenomenon through the differing perspectives between environmental and spatial geography, "institutional economics" and "ecological economics", and even by the gap between public and private law, although these two approaches converge when dealing with an environmental issue.

Although it is impossible for any mediator to approach these disciplines in a straightforward manner, another strategy would be to propose certain "experiential" or "recreational" activities in order to better involve the reader to consider and understand the issues at stake. For example, the reader could be presented with a particular and potentially problematic situation, in the form of a scenario, and be asked to take a stance based on the disciplinary elements at their disposal. This would require them to link the different disciplinary perspectives in order to propose a global solution, integrating a set of viewpoints on a specialized issue, thus putting "specialty" back at the heart of the discussion (Van der Yeught 2019).

This type of scenario was tested during a research seminar with the Valbio collective (in March 2019) with this particular objective. The multidisciplinary actors were presented with a situation of "regulatory disruption", which was set like a political-fiction scenario: from such and such a date (in the near future), chemical pesticides will be banned and biocontrol will be made mandatory (see Box 8.1 for the entire scenario).

Three multidisciplinary teams took charge of the scenario, in order to think about how to structure the sector in the face of this disruption and to propose solutions to the necessary change in practices. After a period of work and exchange between the members of each group, each one presented its recommendations[50] and the whole

49 Original French version: "[…] *loin d'avoir l'unité épistémologique que leur intitulé leur prête, sont des espaces de controverses et rassemblent des orientations de recherche extrêmement diverses qui peuvent être en contradiction les unes avec les autres, et même purement et simplement s'ignorer*" (Jollivet 2008, p. 13).

50 The first group thinks of a conventional organization system into sectors. Biocontrol is then considered as a set of commercial services and products, associated with major actors. The second group thinks about the organization of biocontrol into public and semi-public services for the management of biodiversity and biocontrol issues. The last group imagines a system composed of local biocontrol organizations supported by local networks; in this context, biocontrol is a practice based on local, public, private and farmer interactions and networks.

group discussed the different proposals, which was represented with the help of visual maps[51] (Boutet and Parmentier-Cajaiba 2022).

In Europe, an unprecedented scandal involving general chemical products, and synthetic pesticides in particular, has highlighted the very serious impact of these products. Indeed, it has been demonstrated that significant hormonal imbalances, physical malformation in newborns and a significant increase in occupational agricultural diseases are now correlated to the use of synthetic chemicals. This observation urges drastic action, so as to ensure that agriculture is maintained while ensuring the safety of consumers.

Over the last 20 years, an increase in biological invasion has been observed. They may or may not be linked to global warming, but this also calls for faster changes in preventive measures (prediction of invasions, measures to intercept invasive species and invasion management plans) in European territories. Under pressure from public opinion, rural territories, civil society actions and industrial interests wishing to shift towards biocapitalism, two laws were passed in a hurry:

– Law 2019-558 of May 30, 2019, which bans all chemically synthesized plant protection products (pesticides, fungicides, insecticides, etc.) from sale, use and export in France and Europe as from 2022.

– The Guillaume-Poisson law of July 10, 2019, which makes the use of biocontrol mandatory and encourages the rapid reorganization of agricultural production systems, as well as the reassessment of environmental management.

Box 8.1. *Political fiction scenario in the field of plant protection products (produced by the Research Design Office collective for Valbio)*

The interest of involving a reader – layman, a specialist or a semi-expert – in this type of scenario (potentially in the form of a game) is to impose a disciplinary and interdisciplinary reflexivity, while exchanging with other members on their own vision of the situation. The final objective is to consider a global perspective, while being informed of the different disciplinary issues.

An interesting strategy for these external sources (whose conditions of interaction would have to be specified) would be to propose new scenarios that take the opposing view to popular imagination, which can be restrictive. The scenarios could, for example:

a) assert that biopesticides are also to be banned and ask the participants to propose new alternative methods for plant protection (with a methodical argumentation);

---

51 Maps made by the artist collective *Bureau d'Études* (see: http://fermedelamhotte.fr/bureau.html) for the Arts-Sciences workshop.

b) explain that agrochemical companies are (also) moving towards biocontrol and ask participants to reflect on the issues (including ethical issues) raised by this change (like some of the press statements that mention this paradox)[52];

c) explain that fields treated with natural products (copper, sulfur) can (likewise) be detrimental to their environment[53] and ask the parties to diagnose this and propose measures to protect neighboring fields.

In the same vein, in contrast to "dominant ideas", we can also suggest a reflection on questions of time and the history of agronomic methods, so as to fight against a form of argumentative evidence, which would consider one method replacing another in an almost disruptive manner: from ancestral, natural and positive agriculture (before the industrial era) to negative chemical agriculture, then to innovative connected and natural agriculture. The evolution of practices is undeniably less schematic and straightforward.

These potentially "disparate" points of view, and the associated questioning, would then make it possible to better bring out the complexity associated with each of these questions. Thanks to the opinions and the interaction of the different actors in agroecology, a space for exchange is created in which "environmental controversy is seen as a model of social learning, capable of encouraging a new way of thinking and acting in an uncertain and complex world"[54] (Boillot Grenon 2015).

---

52 "With chemical pesticides, Monsanto and his peers use the same lie. Given that we don't want them anymore, they replace them with 'biocontrol' products. The propaganda shows us some flowers and aphid-eating ladybugs. But behind these beautiful pictures is an army of patented products that have been synthesized in the laboratory, bad genetically modified copies of bacteria, genetic sequences or substances produced by plants" (*L'Humanité*, 2016).

Original French version: "*Avec les pesticides chimiques, Monsanto et ses pairs utilisent le même mensonge. Puisque nous n'en voulons plus, ils les remplacent par des produits de 'biocontrôle'. La propagande nous montre quelques fleurs et des coccinelles mangeuses de pucerons. Mais derrière ces belles photos se cache une armée de produits brevetés synthétisés au laboratoire, mauvaises copies génétiquement modifiées de bactéries, de séquences génétiques ou de substances produites par les plantes*" (*L'Humanité*, 2016).

53 "The hypothesis is rather against prevailing ideas that organic, less-treated fields are reservoirs for pests and pathogens, and instead harm their surroundings" (*Le Monde*, 2016).

Original French verison: "*L'hypothèse est plutôt à contre-courant des idées dominantes qui considèrent que les champs bio, moins traités, sont des réservoirs à ravageurs et à pathogènes et qu'ils portent plutôt préjudice à leur entourage*" (*Le Monde*, 2016).

54 Original French version: "*La controverse environnementale est envisagée comme modèle d'apprentissage social capable de favoriser une nouvelle façon de penser et d'agir dans un monde incertain et complexe*" (Boillot Grenon 2015).

There is another initiative by INRAE in line with these principles of recreational activity for better public involvement: the creation of a collaborative game named "STÅL Invasions" in 2021[55]. This game is set on a new planet with limited resources. Players must develop strategies to protect their crops in order to survive. The game's objective is to be aware of the difficult mission of agroecology's different actors, while always considering the benefit–risk balance involved in this issue of plant protection. All of these playful and experiential initiatives promote the participants' involvement in their own learning, through skills of analysis, dialogue, strategy and risk taking.

Through the necessary global perspective on a subject that allows citizens to take an affirmed position, they enter into a process of reflection and arbitration, while simultaneously mobilizing their own knowledge and convictions. In this way, they take on the position of a researcher, faced with particularly complex interdisciplinary fields.

55 The name of the game comes from a famous 19th century Swedish entomologist, Carl Stål. See: https://www.inrae.fr/actualites/stal-invasions-jeu-familial-developpe-scientifiques-dinrae-sensibiliser-lagroecologie [Accessed March 2, 2022].

Conclusion

This research works investigates the media coverage of scientific subjects in the daily press, using a terminological and cognitive approach. After evoking the challenges inherent to "communicating" science in the media, we present a study of French and English press discourses related to the field of agroecology. The analysis relies on textual elements observed in corpus, presenting both discursive significance in terms of explanation (or even argumentation), as well as limitations from an interpretative standpoint.

The initial phenomenon under observation pertains to the terminological variation involved in the naming and designations of methods, agents and products in this field of specialization. We show how the coexistence of terms, which rely on units that are common in general language (*biological, protection*, *natural, regulation*, etc.), makes it possible to turn complex discourse objects into more familiar ones, therefore making them more accessible to the lay person. However, the analysis shows that this competitive designation process can be detrimental to the comprehension of this field by the non-initiated reader. Based on the assumption that the circulation of terms goes hand-in-hand with the circulation of knowledge (Delavigne 2005), we present a set of cognitive issues that are induced by a proliferation of terms in ordinary language, which result from the determinologization process.

The second phenomenon being studied relates to issues of simplification, regarding both the methods and the reasoning behind the processes involved. On the one hand, we show the recurrence of a binary framing in the description of biocontrol methods, agents and products (opposing biological and chemical methods). On the other hand, we describe the analogical and simplified reasoning used by the media: in particular, metaphors relating to conflict (*biological warfare, guerre biologique*), food (*the wasps' appetites, croquer les pucerons*), cultural or film references (*entomological remake of the film Alien*, *The Ladykillers*), etc.

The advantage of these discursive strategies is that they bring complex phenomena within the reader's reach. However, they may also create interpretative difficulties for a non-specialist, as the latter may not be able to measure the complexity of the processes: for example, the inappropriate opposition between chemical and biological products in certain situations ("*no more chemicals, make way for alternatives*") or the unsuitable pest eradication instead of regulation ("*exterminate aphids*"), etc. Such simplifications may not help the reader to use reflexivity, which is necessary for a proper understanding of the interdisciplinary field under study.

However, the choices of terminology and framing also have a larger interpretative impact: in addition to the implicit scientific and (bio)technological questions, societal, bioethical and legal issues are at stake, particularly on the question of human interference in natural phenomena. This is all the more complex when we consider the European and international context and the relevant legislation. Indeed, as the terminology and regulations concerning pest control agents and biocontrol products are specific to French, European or international institutions, the respective categorizations are, generally speaking, not equivalent.

In an area where the elements of knowledge and language are not stabilized in both the science and legal worlds, the journalist's mission is unquestionably of great complexity. As an actor in this dialogue between science and society, what challenges do they face on these complex questions relating to the environment?

How can we provide the non-specialist with methods and a way of thinking, while presenting specific scientific terminology that is far removed from their own? This mission does not simply require reducing or reformulating specialized terminology, but also "restructuring thought"[1] (Thouin and Lapointe 2016), while also contextualizing this information within societal dynamics and citizens' lives. Journalists thus face two simultaneous challenges: not only do the concepts studied have little in common with what they mean for the public[2], but scientific discourse also relies on terms and expressions, attributing definitions to the latter that are often very different from common meaning (Lamontagne 2016).

This raises a number of questions: is it possible to inform or explain complex processes in a simple way, while targeting a heterogeneous audience? One of the difficulties stems from the fact that this audience already has pre-acquired knowledge. Journalists are faced with a dilemma, having to strike a balance between

1 Original French version: "[…] *de réduire ou de reformuler la terminologie spécialisée, mais également de 'restructurer la pensée'*" (Thouin and Lapointe 2016).

2 The authors quote the example of crystal for a scientist, which is very different from what the concept means for the public.

the reader's knowledge and the specialist knowledge of the subject matter: "On the one hand, less specialized knowledge to target the greatest number, and on the other, more specialized information but a limited readership"[3] (Charaudeau 2011, p. 11). Charaudeau reminds us that the more information is understood by a large number of people, the less it informs (Charaudeau 2011, p. 229). Aspiring to reach the widest audience while effectively informing readers efficiently can seem paradoxical. However, this remains a commendable and necessary objective, assigned to the mediation of science, and which cannot a priori be open to contradiction.

Finally, how do we attract, capture or even captivate a readership without delving too far into the discourse strategies of simplification? Formally, press articles must attract readers by providing information on the subjects involved. The very nature of the media, and of the daily newspaper in particular, is to thrive on what is new, or to bring an original perspective on a situation that is already known (Thouin and Lapointe 2016). Mediators are thus likely to present research results as evidence of an imminent catastrophe or, conversely, as sensational discoveries with stunning applications, like certain journalists in politics, economics and general news articles (Thouin and Lapointe 2016).

All these challenges and obstacles are further exacerbated in the context of a crisis, with the laws of the market to contend with. Cabedoche mentioned the journalism crisis as early as 2003, both internally (in response to market and competitive dynamics) and externally (degradation of the journalist's image) (Cabedoche 2003, p. 227). This tendency has been confirmed over the past two decades, notably by the upheaval of journalistic practices due to the development of digital technologies (Ringoot and Utard 2005; Pélissier and Maas 2014; Lafon 2019b). Addressing scientific issues in this context is now even more difficult, because of the limited time that a journalist has to produce a news story, the reduced space devoted to science in the media and the fact that the newspaper industry is struggling to cope with competition from new communication media.

This competitive framework explains and justifies the strategies used to attract readers (e.g. sensationalism, or the use of socio-cultural stereotypes), even though they can be detrimental to the image of the media. Thouin and Lapointe (2016) speak of a potential "self-destructive tendency", because by scaremongering, there is a risk of losing the public's attention or of transforming the information into a commercial product. However, this process is inherent to the work of the journalist and the press, and for which they can hardly be blamed. As Delforce already

3 Original French version: "*D'un côté un degré de savoir peu élevé, pour viser le plus grand nombre, et de l'autre, une teneur en savoir forte mais une limite en termes de lectorat*" (Charaudeau 2011, p. 11).

expressed: "Blaming the journalist for social stereotypes would be blaming a practitioner for a system"[4] (Delforce 2003, p. 168). The author also emphasizes that the journalist conveys what comes out of successive mediations and that "it is difficult to demand accuracy or correctness of scientific knowledge as in the typical mission of a science educator, which is not their role"[5].

We know that the daily press does not have the same objectives as popular science publications. Jacobi reminds us that there has always been an alternation between two traditional postures in the coverage of science and technology: one that falls within the "tradition of popularization rhetoric proposing a decryption of specialist jargon for laymen; and another which is the description of scientific and technical facts according to the principles of the coverage of current affairs"[6] (Jacobi 2005, p. 64). Our work is more concerned with this second posture. However, the objective of the third part of our book (and this conclusion) is to broaden our perspective by considering the possible convergences between these two types of journalism (specialized and general) and by considering different strategies for communicating both scientific and environmental issues to a less expert public. Throughout the book, we have emphasized the fact that although the media participate in the co-construction of social representations of scientific research, they are in no way the only ones responsible for the dissemination of information, the responsibility being largely shared between the different actors of this process (researchers, mediators, and the public).

The analysis at the heart of this work, as well as our investment in this project since the outset, has led us to the conclusion that there needs to be a closer cooperation between the network of researchers, that of "communicators" and the general public. In addition to the already existing exchanges (seminars, interviews, press releases, etc.), collaboration should be undertaken with new communication models that are adapted to public diversity. An information and exchange platform would bring actors in the process closer together, whether this be a matter of proposing solutions to the variety of terms used, providing reflective elements on the interdisciplinarity and scientific uncertainty characterizing this field (particularly its unintended effects) or involving the public by presenting scenarios that are both

---

4 Original French version: "*Imputer au journaliste la responsabilité des stéréotypes sociaux, ce serait imputer à un acteur ce qui revient à un dispositif*" (Delforce 2003, p. 168).

5 Original French version: "*Placer le journaliste – comme le scientifique intervenant dans les débats publics –, dans la situation typique du didacticien des sciences, qui n'est pourtant pas la sienne*" (Delforce 2003, p. 168).

6 Original French version: "[…] *tradition de la rhétorique de vulgarisation proposant un déchiffrement pour profanes des jargons de spécialistes; et la description des faits scientifiques et techniques selon les principes du traitement de l'actualité*" (Jacobi 2005, p. 64).

entertaining and analytical. Addressing these different issues is of major importance, as they are currently one of the obstacles blocking development in the field of biocontrol.

All the actors in this process would benefit from greater collaboration. Along the lines of Jollivet, we struggle for a more systemic approach in the coverage of environmental issues. The author reminds us that any reflexivity in this field necessarily implies the association of social sciences and natural sciences. Social scientists will discover the "biophysical-chemical aspects" of human action on the environment (Jollivet 2008, p. 17), and natural scientists will be able to better understand that societies are "complex objects" in the image of their own scientific objects. "The phenomena they study are rooted in this complexity. Consequently, they cannot understand them, much less impose their explanations, by isolating them from the latter and simply considering that 'all else being equal'"[7] (Jollivet 2008, p. 17).

Ultimately, the future of this field of study lies in this three-party collaboration between scientists, journalists and citizens. In the development of participatory sciences, these different parties can become actors in this communication, and even in the production of knowledge through critical thinking and involvement into the complexity of the subject. Each of these three parties will also be accompanied in their approach by external sources of qualitative and stable information on the specialized field.

In this study, we tackle the more general issue of communicating on "innovation" which is particularly relevant to the concept of biocontrol. Generally speaking, any innovation challenges existing practices or standards. There are bound to be objections, as well as a potentially negative preconception to its reception. This is a typical trajectory for any innovation (Gaglio 2021). This may include old standards, which we can refer to using the term retro-innovation, or innovative ones, relying on simple technologies, that can be qualified as "low-tech". It can also include disruptive models characterized by innovative practices. This sometimes leads to a change in norms or the creation of new standards, or a reversal of the dominant perspective. It takes time, effort and adaptation to new practices for the situation to become clearer and more stable.

In our case study, not all prejudice is negative from the outset. Biological control is even viewed in a favorable light in the press if we consider its objectives of

7 Original French version: "[...] *que les phénomènes qu'ils étudient s'enracinent dans cette complexité, qu'en conséquence, ils ne peuvent les comprendre et encore moins imposer leurs explications, en les isolant de celle-ci et en la renvoyant dans le 'toutes choses égales par ailleurs*'" (Jollivet 2008, p. 17).

replacing phytosanitary products. However, we are not confronted with "disruptive innovation", for at least two reasons: on the one hand, natural methods remain complementary to chemical methods; on the other hand, there are points of convergence between these new practices and older methods (well before chemical methods), despite the biotechnological processes involved being completely different. However, for innovation that is sensitive in certain aspects, particularly in terms of unintended effects, extra communication is essential to ensure that non-specialists have a better understanding of the processes, objectives and stakes involved.

Consequently, what are the future prospects for this subject in the media? What relevant and realistic objectives can be pursued?

Thouin and Lapointe (2016) push for "open scientific mediation" and compare scientific mediation to art criticism in contemporary art, creating a form of lightness, and also responsibility. In the same way as art criticism aims to elucidate the context and production methods of artworks, "scientific mediation could also set itself the objectives of presenting the specificity of the scientific method, and of providing a reference framework for the social and political choices which involve the technological applications of science"[8]. It is therefore not a matter of educating the citizen, but of involving them in the thought process and of taking a critical look at science and its impact on society (Thouin and Lapointe 2016). This is in line with our initial hypothesis. In order to better understand scientific information, some basic knowledge is required. Fundamental and stable knowledge should be shared among all parties so as to clearly and intelligently contribute to current debates on scientific subjects and, in this case, on the agroecological transition.

Some see it as a kind of "common good" that must be shared, as Caune puts it: in the fields of science, culture or education, knowledge is "a universal heritage, a factor of progress and evolution, a public wealth. Sharing it means pooling information, which manifests itself through transmission: transgenerational, transnational, transcultural transmission"[9] (Caune 2008, p. 41). However, the author specifies that "contrary to the common meaning of the word *sharing*, this pooling is

8 Original French version: "*La médiation scientifique pourrait aussi se fixer les objectifs de présenter la spécificité de la méthode scientifique, et de fournir une grille d'analyse des choix sociaux et politiques qu'impliquent les applications technologiques de la science*" (Thouin and Lapointe 2016).

9 Original French version: "[...] *un patrimoine universel, un facteur de progrès et d'évolution, une richesse publique. Le partager équivaut à une mise en commun qui se manifeste par la transmission: transmission transgénérationnelle, transnationale, transculturelle*" (Caune 2008, p. 41).

not a distribution feature, but a multiplying feature. Sharing is experienced as a source of wealth that increases as it spreads"[10] (Caune 2008, p. 41).

These external information resources for the general public can thus ensure basic knowledge, which then serves as a basis for the acquisition of new knowledge, in order to better understand future developments.

In the current field of investigation, based on the assumption that the next few years will open up new agroecological perspectives, it is impossible do without a certain body of knowledge related to the history of agricultural practices. This concerns, for example, the increase in chemical treatments since the 20th century, associated with "traditional" methods of plant protection; new, potentially hybrid models, compromising high yields with limited use of chemicals and less exploitation of natural resources (sustainable agriculture, ecologically intensive agriculture (EIA), etc.); or other developing agricultural models advocating a more systemic vision, aimed at restoring and maintaining the unity of ecosystems (biodynamics).

As a closing and opening remark, we envisage a new hypothesis on the future media coverage of agroecology. With the development of these alternative methods to chemical treatments, a reversal of both norms and assumptions is likely to happen in press discourse: natural plant protection mechanisms could become "conventional treatments", and chemical methods could be described as "outdated methods", belonging to another era. In any case, a good knowledge of the history of agricultural practices and of the evolution of citizens' perspectives will enable readers to better understand and interpret press discourses on these subjects. It is thus possible to perceive the present and future importance and necessity of the humanities and social sciences, as a whole, in the interpretation of scientific questions. Ultimately, the mediation of science and the dissemination of scientific knowledge raise multi-faceted and multi-disciplinary issues, for which applied linguistics and terminology prove to be a determining factor given their objects of study: language, discourse and also knowledge. They are therefore particularly well suited to providing tools for understanding and reflection in order to improve communication and the circulation of knowledge.

10 Original French version: "[...] *mise en commun n'est pas, contrairement au sens courant du mot partage, une répartition mais au contraire un facteur multiplicateur. Le partage est vécu comme une source de richesse qui s'accroît au fur et à mesure qu'elle se répand*" (Caune 2008, p. 41).

Glossary

This glossary lists the main French terms used in this book, along with their English equivalents and a general definition (adapted from the terminological database of the Vegephyl association)[1].

***Agroécologie* (Agroecology)**: a set of agricultural practices that prioritize biological interactions and aim for the optimal use of possibilities offered by agroecosystems by integrating all available methods, with due consideration for human well-being and environmental integrity.

***Auxiliaire* (Auxiliary)**: a predatory or parasitic animal, that, through its living conditions, contributes to the control of pests detrimental to crops.

Note: generally speaking, auxiliaries are animals, but they can also be micro-organisms, pathogens or even plants (JORF 2018).

See: *parasite* (parasite), *prédateur* (predator).

***Bioagresseur* (Pest)**: a living organism that causes damage to cultivated plants or crops.

Note 1: They can also be plant pathogens or weeds.

Note 2: the term *ennemi des cultures* (*crop enemy*) is also sometimes used in this sense (JORF 2018).

See: *organisme nuisible* (harmful organism), *ravageur* (pest).

---

1 See: https://www.vegephyl.fr/publications/bases-de-donnees/base-repertoire-terminologie/ [Accessed March 7, 2022].

Biocide **(Biocide)**: see *produit biocide* (biocidal product).

Biopesticide **(Biopesticide)**: see: *produit de protection biologique* (biological protection product), *protection biologique* (biological protection).

Bioprotection **(Bioprotection)**: see: *protection biologique* (biological protection).

Lutte biologique **(Biological control)**: see: *protection biologique* (biological protection).

Lutte chimique **(Chemical control)**: see: *protection chimique* (chemical protection).

Lutte intégrée **(Integrated Pest Management)**: the careful consideration of all available plant protection methods and, consequently, the integration of appropriate measures that discourage the development of pest populations, keeping the use of plant protection products and other types of intervention to levels that are economically and environmentally justified, and reduce or minimize risks to human health and the environment. Integrated Pest Management promotes the growth of healthy crops with minimal disruption to agroecosystems, and also encourages natural pest control mechanisms (Dir. 2009/128, see Chapter 4, Note 3).

Synonym: *protection intégrée* (integrated protection).

Organisme nuisible **(Harmful organism)**: a living organism deemed harmful to humans, cultivated plants, domestic animals, or food products, due to the losses it induces or the risks it poses to humans or animals.

See: *bioagresseur* (pest), *ravageur* (pest).

Parasite **(Parasite)**: an animal or plant organism that develops at the expense of a host organism, throughout all or part of its life cycle, causing harm without necessarily causing its death. Endoparasites develop inside the host, while ectoparasites evolve outside the host.

Example: flea, bedbug, tick, mistletoe, broomrape.

Parasitoïde **(Parasitoid)**: an animal or plant organism that develops at the expense of a host, causing its inevitable death. There are two types: endoparasitoids, which develop inside the host, and ectoparasitoids, which develop outside the host.

Example: trichogramma, entomopathogenic fungi.

Pesticide **(Pesticide)**: within the regulations of the European Union, this term encompasses all plant protection products (see this term) and biocidal products (see this term).

Note: the term *pesticide* is often used to designate phytosanitary products.

Phéromone **(Pheromone)**: a chemical mediator that plays a role in the interactions between individuals of the same species by triggering characteristic physiological reactions or behaviors.

Phytopharmacie **(Phytopharmacy)**: the study of active substances and preparations intended for the protection or improvement of plant production, the preservation of harvested products, or the maintenance of non-cultivated areas, excluding fertilizers and amendments.

Synonym: *agropharmacie* (agropharmacy).

See: *produit phytopharmaceutique* (phytopharmaceutical product).

Prédateur **(Predator)**: an animal organism that actively pursues and captures live prey for nourishment or to feed its offspring. Some insects immobilize the prey intended for their larvae.

Example: wasps.

Produit biocide **(Biocidal product)**: a substance or preparation designed to control harmful organisms through chemical or biological action.

Example: products for treating livestock facilities, household insecticides.

See: *organisme nuisible* (harmful organism), *produit phytopharmaceutique* (phytopharmaceutical product).

Note: in practice, this term refers to products that are not human or animal medicines, cosmetics or phytopharmaceutical products.

Produit de protection biologique **(Biological protection product)**: an organism, substance or preparation used to regulate bioaggressor populations through biological protection (see this term).

Note 1: this term should be favored over “biopesticide”.

Note 2: the term “biocontrol product” is also used in everyday language.

Produit phytopharmaceutique **(Phytopharmaceutical product)**: an active substance or preparation intended to:

– protect plants or plant products against harmful organisms or prevent their action;

– act on the vital processes of plants, provided it does not involve nutritive substances;

– ensure the preservation of plant products, insofar as these substances or products are not subject to specific dispositions;

– destroy unwanted plants or parts of plants, as well as unwanted plant growth, with the exception of algae, unless the products are applied to soil or water to protect plants.

Synonym: *produit phytosanitaire* (phytosanitary product), *produit de protection des plantes* (plant protection product).

See: *organisme nuisible* (harmful organism), *produit biocide* (biocidal product).

Note: this term should be favored over "pesticide", which covers phytopharmaceutical products and some biocidal products.

Protection biologique **(Biological protection)**: a method involving the regulation of populations of pests through the use of natural agents that act on natural mechanisms. These agents can be animals (macro or micro-organisms), plants or mineral substances.

Note: this term should be favored over the French term "biocontrôle".

Synonym: *bioprotection* (bioprotection), *lutte biologique* (biological control).

Protection chimique **(Chemical protection)**: the use of chemical phytopharmaceutical products to control harmful organisms.

Synonym: *lutte chimique* (chemical control).

Ravageur **(Pest):** a harmful organism that lives at the expense of plants or commodities, causing damage. When the damage is minor, it is referred to as a *depredator*.

References

ACPM (2016). One 2015–2016. L'audience de la presse en France. Press release [Online]. Available at: https://www.acpm.fr/Media/Files/Communique-ACPM-ONE-ONE-Global [Accessed 22 February 2022].

Albertini, J.-M. and Bélisle, C. (1988). Fonctions de la vulgarisation scientifique. In *Vulgariser la science ou le procès de l'ignorance*, Jacobi, D. and Schiele, B. (eds). Champ Vallon, Seyssel, 223–245.

Allan, S. (2011). Introduction: Science journalism in a digital age. *Journalism*, 12(7), 771–777.

Allard-Huver, F. (2020). Gilles-Éric Séralini ou la transgression des médiations "traditionnelles" du savoir ? *Revue française des sciences de l'information et de la communication*, 20 [Online]. Available at: http://journals.openedition.org/rfsic/9397 [Accessed 22 February 2022].

Allard-Huver, F. (2024). Les contestations de la transition agroécologique : peut-on parler d'agribashing en France ? In *Transitions en tension : controverses et tensions autour des transitions écologiques*, Catellani, A. and Lits, G. (eds). ISTE Editions, London.

Altmanova, J., Grimaldi, C., Zollo, S.-D. (2018). Le rôle de l'adjectif dans la catégorisation des déchets. *6e Congrès Mondial de Linguistique Française*, 46. doi: 10.1051/shsconf/20184605004.

Amend, E. and Secko, D. (2012). In the face of critique: A qualitative meta-synthesis of the experiences of journalists covering health and science. *Science Communication*, 32(2), 241–282.

Amiel, P. (ed.) (2020). Histoire du journalisme de solutions. *Le journalisme de solutions*. PUG, Fontaine, 9–31.

Amiel, P. (2021). Le journalisme de solutions. *Communication*, 34(2) [Online]. Available at: http://journals.openedition.org/communication/7226 [Accessed 22 February 2022].

Attruia, F. (2020). Les discours de l'Union européenne sur la biodiversité : une analyse des registres communicationnels. *Les Carnets du Cediscor*, 15 [Online]. Available at: http://journals.openedition.org/cediscor/2376 [Accessed 22 February 2022].

Aulagnier, A. and Goulet, F. (2017). Des technologies controversées et de leurs alternatives. Le cas des pesticides agricoles en France. *Sociologie du travail*, 59(3) [Online]. Available at: http://journals.openedition.org/sdt/840 [Accessed 22 February 2022].

Aussenac-Gilles, N. and Condamines, A. (2007). Corpus et terminologie. In *La redocumentarisation du monde*, Pédauque, R.T. (ed.). Cépaduès, Toulouse, 131–147.

Authier, J. (1982). La mise en scène de la communication dans des discours de vulgarisation scientifique. In *Langue française*, volume 53, Mortureux, M.F. (ed.). Larousse, Paris, 34–47.

Avenier, M.-J. and Parmentier Cajaiba, A. (2012). The dialogical model: Developing academic knowledge for and from practice. *European Management Review*, 9(4), 199–212.

Aymerich, J.F., Fernández Silva, S., Cabré Castellví, M.T. (2008). La multiplicité des chemins dénominatifs. *Meta*, *53*(4), 731–747.

Babou, I. (2017). L'atelier politique de la nature. *Questions de communication*, 32 [Online]. Available at: http://journals.openedition.org/questionsdecommunication/11423 [Accessed 22 February 2022].

Ballarini, L. (2008). Presse locale, un média de diversion. *Réseaux*, 2(148/149), 405–426.

Baudouin, J. (2009). *Les problèmes théoriques de la vulgarisation scientifique*. Archives Contemporaines/EAN, Paris.

Bauer, M. (1998). The medicalization of science news: From the 'rocket-scalpel' to the 'gene-meteorite' complex. *Social Science Information*, 37(4), 731–751.

Bauer, M. and Shoon, L. (1993). Mapping variety in public understanding of science. *Public Understanding of Science*, 2, 141–155.

Bauer, M., Allum, N., Miller, S. (2007). What can we learn from 25 years of PUS survey research? Liberating and expanding the agenda. *Public Understanding of Science*, 16(1), 79–95.

Beacco, J.-C. (2000). Écritures de la science dans les médias. *Les Carnets du Cediscor*, 6 [Online]. Available at: http://journals.openedition.org/cediscor/319 [Accessed 22 February 2022].

Beacco, J.-C. and Moirand, S. (1995). Autour des discours de transmission des connaissances. *Langages*, 117, 32–53.

Beacco, J.-C., Claudel, C., Doury, M., Petit, G., Reboul-Touré, S. (2002). Science in media and social discourse: New channels of communication, new linguistic forms. *Discourse Studies*, 4(3), 277–300.

Beaudet, C. (2000). La vulgarisation scientifique est-elle possible ? L'affaire Sokal : grandeurs et misères de l'interdisciplinarité. *Technostyle*, 16(1), 135–148.

Bednarek, M. and Caple, H. (2014). Why do news values matter? Towards a new methodological framework for analyzing news discourse in critical discourse analysis and beyond. *Discourse & Society*, 25(2), 135–158.

Bensaude-Vincent, B. (2010). Splendeur et décadence de la vulgarisation scientifique. *Questions de communication*, 368 [Online]. Available at: http://journals.openedition.org/questionsdecommunication/368 [Accessed 22 February 2022].

Bensaude-Vincent, B. (2018). S'approprier ou partager la science ? *La Pensée*, 396(4), 21–31.

Bernard, J.-L. (2017). *Biocontrôle en protection des cultures. Périmètre, succès, freins, espoirs*. L'Harmattan, Paris.

Bhatia, V. (2002). Applied genre analysis: A multi-perspective model. *Ibérica: Revista de la Asociación Europea de Lenguas para Fines Específicos (AELFE)*, 4, 3–19.

Biber, D., Johansson, S., Leech, G., Conrad, S., Finegan, E. (1999). *The Longman Grammar of Spoken and Written English*. Longman, Harlow [Online]. Available at: https://www.degruyter.com/database/COGBIB/entry/cogbib.1245/html [Accessed 19 January 2024].

Boillot Grenon, F. (2015). L'émergence tardive de la communication de l'environnement et du développement durable. *Hermès, La Revue*, 71, 161–163.

Bonnafous, S. (2011). La médiatisation de la question immigrée : état des recherches. *Études de communication*, 22 [Online]. Available at: http://journals.openedition.org/edc/2344 [Accessed 22 February 2022].

Botero, N. (2021). Soustraire l'alimentation du débat sur les pesticides : traitement médiatique d'un problème en une. In *Qu'est-ce que l'on mange ? Les savoirs alimentaires à l'aune des Sciences de l'Information et de la Communication*, Hugol-Gential, C. (ed.). EUD, Dijon, 103–118.

Bouchet, M. (forthcoming). Caractérisation d'un discours numérique spécialisé : le discours administratif de l'accès aux droits. PhD Thesis, Université Paris Cité.

Bourassa, E., Amend, E., Secko, D. (2013). A thematic review and synthesis of best practices in environment journalism. *Journal of Professional Communication*, 3(1), 39–65.

Bourigault, D. and Slodzian, M. (1999). Pour une terminologie textuelle. *Terminologies nouvelles*, 19, 29–32.

Bousquet, F., Marty, E., Smyrnaios, N. (2015). Les nouveaux acteurs en ligne de l'information locale vers une relation aux publics renouvelée ? *Sur le journalisme*, 4(2), 48–61.

Boutet, M. and Parmentier-Cajaiba, A. (2020). Le biocontrôle : perspectives de structuration d'un secteur en développement. In *Biocontrôle – Éléments pour une protection agroécologique des cultures*, Lannou, C., Fauvergue, X., Rusch, A., Barret, M., Jacquin-Joly, E., Malausa, T. (eds). Quae, Versailles, 245–258.

Boutet, M. and Parmentier-Cajaiba, A. (2022). Dialogues entre disciplines, tisser les savoirs et renouveler l'enquête. In *Le Protocole Coopair. Collectif Origens*, Landivar, D. and Monnin, A. (eds). Editions Open Origens, Clermont-Ferrand, 99–114.

Bouyé, M. (2022). Le style clair en anglais de spécialité : caractérisation des schémas lexico-grammaticaux et des ressources interpersonnelles dans deux séries de corpus comparés issus des procédures de demande d'asile dans le droit anglophone et francophone. PhD Thesis, Université de Paris, Paris.

Bowker, L. (1997). You say "flatbed colour scanner", I say "colour flatbed scanner": A descriptive study of the influence of multidimensionality on term formation and use with special reference to the subject field of optical scanning technology. *Terminology. International Journal of Theoretical and Applied Issues in Specialized Communication*, 4(2), 275–302.

Bowker, L. and Meyer, I. (1993). Beyond "textbook" concept systems: Handling multidimensionality in a new generation of term banks. In *Proceedings of the 3rd International Congress on Terminology and Knowledge Engineering (TKE'93)*, Indeks Verlag, Frankfurt, 123–137.

Brossard, D. and Lewenstein, B. (2010). A critical appraisal of models of public understanding of science: Using practise to inform theory. In *Communicating Science: New Agendas in Communication*, Kahlor, L.A. and Stout, P. (eds). Routledge, New York, 11–39.

Bubela, T., Nisbet, M.C., Borchelt, R., Brunger F., Critchley, C., Einsiedel, E., Geller, G., Gupta, A., Hampel, J., Hyde-Lay, R. et al. (2009). Science communication reconsidered. *Nature Biotechnology*, 27(6), 514–518.

Bucchi, M. (1996). When scientists turn to the public: Alternative routes in science communication. *Public Understanding of Science*, 5, 375–394.

Bucchi, M. (2016). Editorial. *Public Understanding of Science*, 25(3), 264–268.

Burns, T.-W., O'Connor, D.-J., Stocklmayer, S.-M. (2003). Science communication: A contemporary definition. *Public Understanding of Science*, 12(2), 183–202.

Cabedoche, B. (2003). Le journaliste, acteur disqualifié de la médiation de l'information scientifique et technique ? In *Communiquer l'information scientifique : éthique du journalisme et stratégies des organisations*, Le Boeuf, C. and Pélissier, N. (eds). L'Harmattan, Paris, 213–250.

Cabré, M.-T. (1998). *Terminology: Theory, Methods and Applications*. John Benjamins, Philadelphia.

Calabrese, L. (2014). Rectifier le discours d'information médiatique. Quelle légitimité pour le discours profane dans la presse d'information en ligne ? *Les Carnets du Cediscor*, 12 [Online]. Available at: http://journals.openedition.org/cediscor/916 [Accessed 25 February 2022].

Calsamiglia, H. (2003). Popularization discourse. *Discourse Studies*, 5(2), 139–146.

Calsamiglia, H. and Van Dijk, T.-A. (2004). Popularization discourse and knowledge about the genome. *Discourse & Society*, 15(4), 369–389.

Candel, D. (2017). Normes en terminologie officielle (France, 1996–2014). *Cahiers de Lexicologie*, 110(1), 29–43.

Carson, R. (1962). *Silent Spring*. Penguin, London.

Catellani, A., Pascual Espuny, C.L., Pudens Malibabo, L., Jalenques Vigouroux, B. (2019). Les recherches en communication environnementale. *Communication*, 36(2) [Online]. Available at: http://journals.openedition.org/communication/10559 [Accessed 22 February 2022].

Catellani, A., Pascual Espuny, C., Jalenques Vigouroux, B. (2021). Les sciences participatives au prisme des Sciences de l'information et de la communication. *Études de communication*, 56 [Online]. Available at: https://journals.openedition.org/edc/11315 [Accessed 22 February 2022].

Caulfield, T. (2004). Biotechnology and the popular press: Hype and the selling of science. *Trends in Biotechnology*, 22(7), 33–79.

Caune, J. (2008). La culture scientifique : une médiation entre sciences et société. *Lien social et Politiques*, 60, 37–48.

Charaudeau, P. (2004). Le contrat de communication dans une perspective langagière : contraintes psychosociales et contraintes discursives. In *Psychologie sociale et communication*, Bromberg, M. and Trognon, A. (eds). Dunod, Paris, 109–120.

Charaudeau, P. (ed.) (2008). *La médiatisation de la science. Clonage, OGM, manipulations génétiques*. De Boeck Supérieur, Brussels.

Charaudeau, P. (2011). *Les médias et l'information. L'impossible transparence du discours*. De Boeck Supérieur, Brussels.

Chateauraynaud, F. and Debaz, J. (2017). *Aux bords de l'irréversible. Sociologie pragmatique des transformations*. Pétra, Paris.

Chateauraynaud, F. and Torny, D. (1999). *Les sombres précurseurs, une sociologie pragmatique de l'alerte et du risque*. Éditions de l'EHESS, Paris.

de Cheveigné, S. (1997). La science médiatisée : les contradictions des scientifiques. *Revue Hermès*, 21, 121–133.

Chlous, F., Dozières, A., Guillaud, D., Legrand, L. (2017). Des recherches participatives dans la production des savoirs liés à l'environnement – Introduction. Foisonnement participatif : des questionnements communs ? *Natures Sciences Sociétés*, 25(4), 327–335.

Ciapuscio, G. (2003). Formulation and reformulation procedures in verbal interactions between experts and (semi)laypersons. *Discourse Studies*, 5(2), 207–233.

Cislaru, G. and Nyckees, V. (2019). *Le partage du sens. Approches linguistiques du sens commun*. ISTE Editions, London.

Classements de presse payante (2022). Alliance pour les chiffres de la presse et des médias (ACPM) [Online]. Available at: https://www.acpm.fr/Les-chiffres/Diffusion-Presse/Presse-Payante [Accessed February 2021].

Clerc, I. (2019). Quelles règles d'écriture se donner pour communiquer avec l'ensemble des citoyens du Québec ? *Éla. Études de linguistique appliquée*, 195, 305–324. doi: 10.3917/ela.195.0305.

Cloître, M. and Shinn, T. (1985). Expository practice. In *Expository Science: Forms and Functions of Popularisation*, Shinn, T. and Whitley, R.D. (eds). Springer, Dordrecht. doi: 10.1007/978-94-009-5239-3_2.

Condamines, A. (1995). Terminology: New needs, new perspectives. *Terminology. International Journal of Theoretical and Applied Issues in Specialized Communication.* John Benjamins, Philadelphia.

Condamines, A. (1997). Langue spécialisée ou discours spécialisé ? In *Mélanges de linguistique offerts à Rostislav Kocourek*, Lapierre, L., Oore, I., Runte, H.R. (eds). Les presses d'Alfa, Halifax, 171–184.

Condamines, A. (2005). Linguistique de corpus et terminologie. *Langages*, Armand Colin, 36–47.

Condamines, A. and Dehaut, N. (2011). Mise en œuvre des méthodes de la linguistique de corpus pour étudier les termes en situation d'innovation disciplinaire : le cas de l'exobiologie. *Meta : journal des traducteurs*, 56(2), 266–283.

Condamines, A. and Picton, A. (2014). Des communiqués de presse du Cnes à la presse généraliste. Vers un observatoire de la diffusion des termes. In *La néologie en langue de spécialité*, Dury, P., Carlos de Hoyos, J., Makri-Morel, J., Maniez, F., Renner, V., Belen Villar Diaz, M. (eds). CRTT, Lyon, 165–188.

Condamines, A. and Rebeyrolle, J. (1996). Point de vue en langue spécialisée. *Meta : le journal des traducteurs*, 42(1), 174–184.

Condamines, A., Dehaut, N., Picton, A. (2012). Rôle du temps et de la pluridisciplinarité dans la néologie sémantique en contexte scientifique. Études outillées en corpus. In *Cahiers de lexicologie. Néologie sémantique et analyse de corpus*, volume 101, Gérard, C. and Katabek, J. (eds). Garnier Classiques, Paris, 161–184.

Couvet, D. and Teyssèdre, A. (2013). Sciences participatives et biodiversité : de l'exploration à la transformation des socio-écosystèmes. *Cahiers des Amériques latines*, 72/73, 49–64.

Crump, N.S., Cother, E.J., Ash, G.J. (1999). Clarifying the nomenclature in microbial weed control. *Biocontrol Science and Technology*, 9, 89–97.

Cruse, D.A. (1986). *Lexical Semantics.* Cambridge University Press, Cambridge.

Daille, B., Habert, B., Jacquemin, C., Royauté, J. (1996). Empirical observation of term variations and principles for their description. *Terminology. International Journal of Theoretical and Applied Issues in Specialized Communication*, 3(2), 197–257.

Davies, S. (2008). Constructing communication: Talking to scientists about talking to the public. *Science Communication*, 29(4), 413–434.

Deguine, J.-P. and Ledouble, H. (2022). Le biocontrôle en France : de quoi parle-t-on ? *Cahiers Agricultures*, 31(19). doi : 10.1051/cagri/2022018.

Deguine, J.-P., Gloanec, C., Laurent, P., Ratnadass, A., Aubertot, J.-N. (2016). *Protection agroécologique des cultures*. Quae, Versailles.

Delavigne, V. (2003). Quand le terme entre en vulgarisation. *Actes des cinquièmes rencontres. Terminologie et intelligence artificielle*, Strasbourg, 31 March and 1 April, 80–91.

Delavigne, V. (2005). Les mots de l'autre : approche contrastive de discours de vulgarisation. In *Didactique du lexique : langue, cognition, discours*, Grossmann, F., Paveau, M.-A., Petit, G. (eds). Ellug, Grenoble, 189–213.

Delavigne, V. (2015). Quand le linguiste devient acteur : vulgarisation médicale et prescription linguistique. In *Cultures de recherche en linguistique appliquée*, Caron, F., Narcy-Combes, J.-P., Narcy-Combes, M.-F., Toffoli, D. (eds). Riveneuve, Paris, 77–95.

Delforce, B. (2003). Les spécificités de la médiatisation : l'information scientifique au risque des médias. In *Communiquer l'information scientifique : éthique du journalisme et stratégies des organisations*, Le Boeuf, C. and Pélissier, N. (eds). L'Harmattan, Paris, 63–175.

Delforce, B. and Estienne, Y. (2011). Le journalisme numérique : formes, formats, frontières. *Les Cahiers du Journalisme*, 22/23.

Dentzer, S. (2009). Communicating medical news – Pitfalls of health care journalism. *New England Journal of Medicine*, 360(1), 1–3.

DeVries, W.C. (1988). The physician, the media, and the 'spectacular' case. *JAMA*, 259(6), 886–890.

Domenec, F. (2020). Aspects rhétoriques et degrés de spécialisation : la communication scientifique, médiatique et institutionnelle autour de l'Affaire Séralini. In *Stratégies et outils rhétoriques dans les discours spécialisés*, Domenec, F. and Resche, C. (eds). Peter Lang, Bern.

Drouin, P., Francoeur, A., Humbley, J., Picton, A. (eds) (2017). *Multiple Perspectives on Terminological Variation*. John Benjamin, Amsterdam.

Dunwoody, S. (2004). How valuable is formal science training to science journalists? *Comunicação e Sociedade*, 6, 75–87.

Dunwoody, S., Hendriks, F., Massarani, L., Hans, P. (2018). How journalists deal with scientific uncertainty and what that means for the audience [conference report]. *15th International Public Communication of Science and Technology Conference*, Dunedin, 3–6 April.

Dury, P. (2006). La dimension diachronique en terminologie et en traduction spécialisée : le cas de l'écologie. In *Aspects diachroniques du vocabulaire*, Gaudin, G. and Candel, D. (eds). Université de Rouen, Université du Havre, 109–124.

Dury, P. (2008). The rise of carbon neutral and compensation carbone: A diachronic investigation into the migration of vocabulary from the language of ecology to newspaper language and vice versa. *Terminology*, 14(2), 230–248.

Dury, P. and Lervad, S. (2008). La variation synonymique dans la terminologie de l'énergie : approches synchronique et diachronique, deux études de cas. *LSP and Professional Communication*, 8(2), 66–79.

Eilenberg, J., Hajek, A., Lomer, C. (2001). Suggestions for unifying the terminology in biological control. *BioControl*, 46, 387–400.

Einsiedel, E. and Thorne, B. (1999). Public responses to uncertainty. *Communicating Uncertainty: Media Coverage of New and Controversial Science*, 43–58.

Entman, R.M. (1993). Framing: Toward clarification of a fractured paradigm. *Journal of Communication*, 43(4), 51–58.

Faber, P., Leon Arauz, P., Prieto Velasco, J.A. (2009). Semantic relations, dynamicity, and terminological knowledge bases. *Current Issues in Language Studies*, 1(1), 1–23.

FAO (2006). Glossary of phytosanitary terms. International Standards for Phytosanitary Measures (ISPM) No 5. International Plant Protection Convention [Online]. Available at: https://www.ippc.int/largefiles/adopted_ISPMs_previousversions/en/ISPM_05_2006_En_2006-07-04.pdf [Accessed 2 September 2022]

FAO (2019). The 10 elements of agroecology: Guiding the transition to sustainable food and agricultural systems. Rome [Online]. Available at: http://www.fao.org/3/na160en/na160en.pdf [Accessed 22 February 2022].

Fauvergue, X., Rusch, A., Barret, M., Bardin, M., Jacquin-Joly, E., Malausa, T. Lannou, C. (2020). *Biocontrôle : éléments pour une protection agroécologique des cultures*. Quae, Versailles.

Fayard, P. (1988). *La communication scientifique publique. De la vulgarisation à la médiatisation*. Chronique sociale, Lyon.

Fernández-Silva, S. (2016). The cognitive and rhetorical role of term variation and its contribution to knowledge construction in research articles. *Terminology*, 22(1), 52–79.

Fernández-Silva, S., Freixa, J., Cabre, T. (2014). A method for analysing the dynamics of naming from a monolingual and multilingual perspective. In *Dynamics and Terminology: An Interdisciplinary Perspective on Monolingual and Multilingual Culture-Bound Communication*, Temmerman, R. and van Campenhoudt, M. (eds). John Benjamins, Amsterdam, 183–211.

Fernbach, N. (2010). La simplification de la langue du droit en contexte anglo-saxon In *Qu'en est-il de la simplification du droit ?* Rueda, F. and Pousson-Petit, J. (eds). Presses de l'Université Toulouse Capitole [Online]. Available at: http://books.openedition.org/putc/1314 [Accessed 22 February 2022].

Fillmore, C.J. and Baker, C.F. (2009). A frames approach to semantic analysis. In *The Oxford Handbook of Linguistic Analysis*, Bernd, H. and Heiko, N. (eds). Oxford University Press, 313–340.

Flaysakier, J.D. (1997). Santé publique et responsabilité des médias. *Revue Hermès*, 21, 135–143.

Foucart, S. (2021). *Et le monde devint silencieux. Comment l'agrochimie a détruit les insectes*. Éditions Points, Paris.

Fouquier, E. and Véron, E. (1986). *Les spectacles scientifiques télévisés : figure de la production et de la réception*. La Documentation française, Paris.

François, T., Müller, A., Degryse, B., Fairon, C. (2018). AMesure, une plateforme web pour soutenir la rédaction simple de textes administratifs. *Repères-Dorif*, 16 [Online]. Available at: http://www.dorif.it/ezine/ezine_articles.php?id=403 [Accessed 22 February 2022].

Freixa, J. (2006). Causes of denominative variation in terminology: A typology proposal. *Terminology*, 12(1), 51–77.

Frewer, L., Howard, J., Shepherd, R. (1997). Public concerns in the United Kingdom about general and specific applications of genetic engineering: Risk, benefit, and ethics. *Science, Technology, and Human Values*, 22(1), 98–124.

Friedman, S.M., Dunwoody, S., Rogers, C.L. (eds) (1999). *Communicating Uncertainty: Media Coverage of New and Controversial Science*. Lawrence Erlbaum, Mahwah.

Gaglio, G. (2021). *Sociologie de l'innovation*. PUF, Paris.

Galochet, M., Longuepee, J., Morel, V., Petit, O. (eds) (2008). *L'environnement : discours et pratiques interdisciplinaires*. Artois Presses Université, Arras.

Gambier, Y. (1991). Travail et vocabulaire spécialisés : prolégomènes à une socioterminologie. *Meta : Journal des traducteurs*, 36(1), 31–58.

Gamson, W.A. and Modigliani, A. (1989). Media discourse and public opinion on nuclear power: A constructionist approach. *American Journal of Sociology*, 95(1), 1–37. doi : 10.1086/229213.

Gardien, E. (2017). Qu'apportent les savoirs expérientiels à la recherche en sciences humaines et sociales ? *Vie sociale*, 4(20), 31–44.

Garric, N. (2018). La définition : construction d'une catégorie (linguistique ? discursive ?). *Semiotica*, 111–126.

Gaudin, F. (1993). *Pour une socioterminologie : des problèmes sémantiques aux pratiques institutionnelles*. Université de Rouen, Mont Saint Aignan.

Gaudin, F. and Alexandru, M.C. (2006). Les contextes : à la source du terme ? *Mots, termes et contextes*, 59–67.

Geeraerts, D., Grondelaers, S., Bakema, P. (1994). *The Structure of Lexical Variation: Meaning, Naming, and Context*. Mouton de Gruyter, Berlin.

Gilbert, C. and Henry, E. (2012). La définition des problèmes publics : entre publicité et discrétion. *Revue française de sociologie*, 53, 35–59.

Gledhill, C., Martikainen, H., Mestivier, A., Zimina, M. (2019). Towards a linguistic definition of 'simplified medical english': Applying textometric analysis to Cochrane medical abstracts and their plain language versions. *Lingue Culture Mediazioni/ Languages Cultures Mediation*, 91–114.

Goffman, E. (1991). *Les cadres de l'expérience*. Les Éditions de Minuit, Paris.

Grize, J.-B. (1990). *Logique et langage*. Ophrys, Paris.

Grize, J.-B. (1992). Éclairage. In *Un signe parmi d'autres*. Gilles Attinger, Hauterive, 22–25.

Grize, J.-B. (2004). Le point de vue de la logique naturelle : démontrer, prouver, argumenter. In *L'argumentation aujourd'hui : positions théoriques en confrontation*, Doury, M. and Moirand, S. (eds). Presses Sorbonne Nouvelle, Paris [Online]. Available at: http://books.openedition.org/psn/758 [Accessed 22 February 2022].

Guespin, L. (1991). La circulation terminologique et les rapports, science technique, production. *Cahiers de linguistique sociale*, 18, 59–80.

Guilbert, L. (1975). *La créativité lexicale*. Larousse, Paris.

Gülich, E. (2003). Conversational techniques used in transferring knowledge between medical experts and non-experts. *Discourse Studies*, 5(2), 235–263.

Halliday, M. (1985). Part A. In *Language, Context, and Text: Aspects of Language in a Social-Semiotic Perspective*, Halliday, M.A.K. and Hasan, R. (eds). Deakin University Press, Geelong, 3–49.

Heiden, S., Magué, J.-P., Pincemin, B. (2010). TXM : une plateforme logicielle open-source pour la textométrie – conception et développement. In *Proceedings of 10th International Conference on the Statistical Analysis of Textual Data – JADT)*, volume 2, Bolasco, S., Chiari, I., Giuliano, L. (eds). Edizioni Universitarie di Lettere Economia Diritto, Rome, 1021–1032.

HLPE (2019). Other innovative approaches for sustainable agriculture and food systems that enhance food security and nutrition. Report, Rome [Online]. Available at: http://www.fao.org/3/ca5602en/ca5602en.pdf [Accessed 22 February 2022].

Holmgreen, L.-L. (2008). Biotech as 'biothreat'? Metaphorical constructions in discourse. *Discourse & Society*, 19(1), 99–119.

Holzem, M. (1999). *Terminologie et documentation. Pour une meilleure circulation des savoirs*. ADBS, Paris.

Holzem, M. and Labiche, J. (2017). *Dessillement numérique : énaction, interprétation, connaissances*. Peter Lang, Brussels.

Houllier, F. and Merilhou-Goudard, J.-B. (2016). Les sciences participatives en France : états des lieux, bonnes pratiques et recommandations [Online]. Available at: https://hal.archives-ouvertes.fr/hal-02801940/ [Accessed 25 February 2022].

Humbert-Droz, J. (2021). Définir la déterminologisation : approche outillée en corpus comparable dans le domaine de la physique des particules. PhD Thesis, Université de Genève [Online]. Available at: https://archive-ouverte.unige.ch/unige:157351 [Accessed 22 February 2022].

Humbley, J. (2011). Vers une méthode de terminologie retrospective. *Langages*, 183(3), 51–62. doi: 10.3917/lang.183.0051.

Humbley, J. (2018). *La néologie terminologique*. Lambert-Lucas, Limoges.

Iyengar, S. (1991). *Is Anyone Responsible? How Television Frames Political Issues*. University of Chicago Press, Chicago.

Jackiewicz, A. and Pengam, M. (2020). Un modèle pour l'étude des nominations émergentes. Notion de repérage pour saisir les modalités d'ajustement sémantique et discursif. *7e Congrès Mondial de Linguistique Française*, Montpellier [Online] Available at: https://www.shs-conferences.org/articles/shsconf/abs/2020/06/contents/contents.html [Accessed 25 February 2022].

Jacobi, D. (1986). *Diffusion et vulgarisation, itinéraires du texte scientifique*. Presses Universitaires de Franche-Comté, Besançon.

Jacobi, D. (2005). Ce qu'il faut de culture (scientifique) pour lire le journal quotidien. In *La publicisation de la science ; exposer, communiquer, débattre, publier, vulgariser*, Pailliart, I. (ed.). PUG, 53–74.

Jacquet-Pfau, C. (2021). La vulgarisation dans les dictionnaires et encyclopédies. Avant-propos. *La linguistique*, 57(1), 3–14.

Jeanneret, Y. (1994). *Écrire la science. Formes et enjeux de la vulgarisation*. PUF, Paris.

Jeanneret, Y. (2000). Des médias, des sciences et des textes. *Les Carnets du Cediscor*, 6 [Online]. Available at: http://journals.openedition.org/cediscor/335 [Accessed 22 February 2022].

Jeanneret, Y. (2004). Le partage des savoirs entre métamorphose des médias et poétique des discours. In *Médiation et représentation des savoirs*, Metzger J.-P. (ed.). L'Harmattan, Paris, 15–32.

Jensen, P., Rouquier, J.-B., Kreimer, O. Croissant, Y. (2008). Scientists connected with society are more active academically. *Science and Public Policy*, 35(7), 527–541.

Jollivet, M. (2007). Les rapports entre sciences et société en question au CNRS : un (faux ?) départ. *Natures Sciences Sociétés*, 15(4), 417–423.

Jollivet, M. (2008). Préface. In *L'environnement : discours et pratiques interdisciplinaires*, Galochet, M., Longuépée, J., Morel, V., Petit, O. (eds). Artois Presses Université, Arras, 11–20.

Jones, S. (2002). *Antonymy: A Corpus-based Approach*. Routledge, London.

Journal Officiel de l'Union Européenne (2009). Directive 2009/128/CE du Parlement Européen et du Conseil, 21 October [Online]. Available at: https://eur-lex.europa.eu/legalcontent/FR/TXT/HTML/?uri=CELEX:32009L0128 [Accessed 22 February 2022].

Journal Officiel de la République Française (2014). Loi n° 2014–1170 du 13 octobre 2014 d'avenir pour l'agriculture, l'alimentation et la forêt [Online]. Available at: https://www.legifrance.gouv.fr/eli/loi/2014/10/13/2014-1170/jo/article_55 [Accessed 22 February 2022].

Jouzel, J.-N. (2019). *Pesticides. Comment ignorer ce que l'on sait*. Presses de Sciences Po, Paris.

Jurdant, B. (1973). Les problèmes théoriques de la vulgarisation scientifique. PhD Thesis, Université de Strasbourg.

Kageura, K. (2002). *The Dynamics of Terminology: A Descriptive Theory of Term Formation and Terminological Growth*. John Benjamins, Amsterdam/Philadelphia.

Kampourakis, K. and McCain, K. (2019). *Uncertainty: How it Makes Science Advance*. Oxford University Press, New York.

Kleiber, G. (1979). À propos de l'ambiguïté Transparence/Opacité. *Travaux de Linguistique et de Littérature*, 17, 1, 233–250.

Kleiber, G. (1984). Dénomination et relations dénominatives. *Langages*, 76, 77–94.

Kleiber, G. (2008). Comment fonctionne ICI. *Cahiers Chronos*, 20, 113–145.

Klein, E. (2020a). *Psychisme ascensionnel*. Arthaud, Paris.

Klein, E. (2020b). *Le gout du vrai*. Gallimard, Paris.

Kovecses, Z. (2005) *Metaphor in Culture: Universality and Variation*. Cambridge University Press, Cambridge/New York.

Krieg-Planque, A. (2009). À propos des "noms propres d'événements". Événementialité et discursivité. *Les Carnets du Cediscor*, 11, 77–90.

Kristiansen, M. and Andersen, G. (2012). Corpus approaches to neology and their relevance for dynamic domains. *Neologica*, 6, 43–62. doi: 0.15122/isbn.978-2-8124-4232-2.p.0043.

Labasse, B. (2012). Sexe, sang et physique des particules : le "sensationnalisme" est-il partout… ou nulle part ? *Les Cahiers du journalisme*, 24, 114–149.

Laberge, Y. (1997). Science, experts, technologie : analyse qualitative du discours de la publicité institutionnelle télévisée. *Revue Hermès*, 21, 203–210.

Lafon, B. (2019a). Des médiatisations au processus de médiatisation. In *Médias et médiatisation. Analyser les médias imprimés, audiovisuels, numériques*. PUG, Fontaine, 157–189.

Lafon, B. (ed.) (2019b). *Médias et médiatisation. Analyser les médias imprimés, audiovisuels, numériques*. PUG, Fontaine.

Lakoff, G. and Johnson, M. (1980). *Metaphors We Live by*. University of Chicago Press, Chicago.

Lambert, E. (2020). Le contentieux américain des victimes de l'exposition au glyphosate. *Revue juridique de l'environnement*, special issue, 201–213

Lamontagne, R. (2016). Pourquoi faudrait-il embaucher plus de journalistes scientifiques ? Tribune. *ACFAS Magazine* [Online]. Available at: https://www.acfas.ca/publications/magazine/2016/12/pourquoi-faudrait-il-embaucher-plus-journalistes-scientifiques [Accessed 22 February 2022].

Laügt, O. (2001). Risque et médiation de la science. *Communication et organisation*, 20 [Online]. Available at: http://journals.openedition.org/communicationorganisation/2578 [Accessed 22 February 2022].

Leach, J., Yates, S., Scanlon, E. (2009). Models of science communication. In *Investigating Science Communication in the Information Age: Implications for Public Engagement and Popular Media*, Holliman, R., Whitelegg, E., Scanlon, E., Smidt, S., Thomas, J. (eds). Oxford University Press, New York.

Ledouble, H. (2019) Vulgarisation scientifique et mediatisation de la science : instabilité terminologique dans le domaine de la lutte biologique. *Terminology. International Journal of Theoretical and Applied Issues in Specialized Communication*. John Benjamins Publishing, 25(1), 60–92.

Ledouble, H. (2020). Term circulation and conceptual instability in the mediation of science: Binary framing of the notions of biological versus chemical pesticides. *Discourse & Communication*, 14(5), 466–488.

Ledouble, H. (2021). Contextes et connaissances dans les discours de vulgarisation scientifique: dynamiques definitoires et problématiques cognitives. In *Des corpus numériques à l'analyse linguistique en langues de spécialité*, Frérot, C. and Pecman, M. (eds). UGA, Grenoble.

Ledouble, H. and Cadel, P. (2022). De la définition juridique à la définition journalistique : complexités interprétatives dans le domaine du biocon-trôle. *Fachsprache*, 44(1/2), 4–19. doi: 10.24989/fs.v44i1-2.1914.

Lefeuvre, L. and Condamine, A. (2015). Constitution d'une base bilingue de marqueurs de relations conceptuelles pour l'élaboration de ressources termino-ontologiques. *Actes de la conférence Terminologie et Intelligence Artificielle*, Granada, 173–178.

Le Marec, J. and Babou, I. (2004). Sciences et médias : le champ "STS" à l'épreuve de la banalité. *Sciences, Médias et Société*. ENS LSH, Lyon [Online]. Available at: http://sciences-medias.ens-lsh.fr/article.php3 ?id_article=54 [Accessed 22 February 2022].

Lerat, P. (1995). *Les langues spécialisées*. PUF, Paris.

Lévy-Leblond, J.-M. (1992). About misunderstandings about misunderstandings. *Public Understanding of Science*, 1(1), 17–21.

L'Homme, M.-C. (2004). *La terminologie : principes et techniques*. Presses de l'Université de Montréal, Montreal [Online]. Available at: http://books.openedition.org/pum/10693. doi: 10.4000/books.pum.10693.

L'Homme, M.-C. and Marshman, E. (2006). Terminological relationships and corpus-based methods for discovering them: An assessment for terminographers. In *Lexicography, Terminology, and Translation. Text-based Studies in Honour of Ingrid Meyer*, Bowker, L. (ed.). University of Ottawa Press, 67–80.

Libaert, T. (ed.) (2016). *La communication environnementale*. CNRS, Paris. doi: 10.4000/books.editionscnrs.20832 [Accessed 2 February 2023].

Lieutenant-Gosselin, M., Brin, C., Fleury, J.-M. (2012). Introduction – Le journalisme scientifique : défis et redéfinition. *Les Cahiers du Journalisme*, 24, 2–13.

Lipani, M.-C. and Pascal, C. (2020). Quand la médiatisation de savoirs "savants" et des chercheur.e.s interrogent le sens de la science. *Revue française des sciences de l'information et de la communication*, 20 [Online]. Available at: http://journals.openedition.org/rfsic/9333. doi: https://doi.org/10.4000/rfsic.9333 [Accessed 22 February 2022].

Loffler-Laurian, A.-M. (1983). Typologie des discours scientifiques : deux approches. *Études de linguistique appliquée*, 51, 8–20.

Loffler-Laurian, A.-M. (1984). Vulgarisation scientifique : formulation, reformulation, traduction. *Langue française*, 64, 109–125.

Logan, R.A. (2001). Science mass communication: Its conceptual history. *Science Communication*, 23(2), 135–163.

Martinez, C. (2021). La traductibilité de réfugiés environnementaux dans le contexte français. *Academic Journal of Modern Philology*, 13(special edition), 203–214.

Marty, E. (2015). Les élections municipales au miroir de la presse quotidienne régionale. Des cadres médiatiques aux thématiques politiques. *Mots. Les langages du politique*, 108, 39–55.

Meilliez, F. (2008). Complémentarité entre les sciences de la matière et du vivant et les SHS : un témoignage d'acteur. In *L'environnement : discours et pratiques interdisciplinaires*, Galochet, M., Longuépée, J., Morel, V., Petit, O. (eds). Artois Presses Université, Arras, 173–178.

Mel'čuk, I., Clas, A., Polguere, A. (1995). *Introduction à la lexicologie explicative et combinatoire*. Duculot, Louvain-la-Neuve.

Meyer, I. (2001). Extracting knowledge-rich contexts for terminography: A conceptual and methodological framework. *Recent Advances in Computational Terminology*, 279–302.

Meyer, I. and Mackintosh, K. (2000). When terms move into our everyday lives: An overview of de-terminologization. *Terminology*, 6(1), 111–118.

Meyer, I., Skuce, D., Bowker, L., Eck, K. (1992). Towards a new generation of terminological resources: An experiment in building a terminological knowledge base. *Proceedings of the 14th Conference on Computational Linguistics*, 3, 956–960.

Miège, B. (2004). *L'information – communication, objet de connaissance*. INA/De Boeck, Brussels.

Miège, B. (2019). Pour une actualisation de l'approche de la médiatisation des actions info – communicationnelles. In *Entre ce qui est dit et ce que vous pensez : où est la médiatisation ?*, Miège, B. (ed.). Universidade Federal de Santa Maria, 33–51.

Miller, S. (2001). Public understanding of science at the crossroads. *Public Understanding of Science*, 10, 115–120.

Millerand, F. (2021). La participation citoyenne dans les sciences participatives : formes et figures d'engagement. *Études de communication*, 56 [Online]. Available at: http://journals.openedition.org/edc/11360. doi: 10.4000/edc.11360.

Moirand, S. (1997). Formes discursives de la diffusion des savoirs dans les médias. *Hermès, La Revue*, 21, 33–44.

Moirand, S. (1999). Éléments de théorisation d'une linguistique du discours : l'exemple du discours sur les sciences dans les médias. *Modèles linguistiques* [Online]. Available at: http://journals.openedition.org/ml/1400 [Accessed 22 February 2022].

Moirand, S. (2000). Variations discursives dans deux situations contrastées de la presse ordinaire. *Les Carnets du Cediscor*, 6 [Online]. Available at: http://journals.openedition.org/cediscor/337 [Accessed 22 February 2022].

Moirand, S. (2003). Communicative and cognitive dimensions of discourse on science in the French mass media. *Discourse Studies*, 5(2), 175–206.

Moirand, S. (2004). De la médiation à la médiatisation des faits scientifiques et techniques : où en est l'analyse du discours ? *Sciences, Médias et Société*, ENS-LSH, Lyon, 15–17 June [Online]. Available at: http://sciences-medias.ens-lyon.fr/article.php3?id_article=59 [Accessed 22 February 2022].

Moirand, S. (2007). *Les discours de la presse quotidienne. Observer, analyser, comprendre*. PUF, Paris.

Moirand, S. (2009). Variations discursives dans deux situations contrastées de la presse ordinaire. *Les Carnets du Cediscor*, 6 [Online]. Available at: http://journals.openedition.org/cediscor/337 [Accessed 22 February 2022].

Moirand, S. (2014). Vers de nouvelles configurations discursives. *Les Carnets du Cediscor*, 12, 141–149.

Moirand, S., Bouacha, A., Beacco, J.-C., Collinot, A. (eds) (1995). *Parcours linguistiques de discours spécialisés*. Peter Lang, Bern.

Moirand, S., Pordeus-Ribeiro, M., Reboul-Touré, S. (2016). La vulgarisation scientifique au croisement de nouvelles sphères d'activité langagière. *Bakhtiniana, Rev. Estud. Discurso*, 11(2). doi: 10.1590/2176-45732387 [Accessed 22 February 2022].

Molek-Kozakowska, K. (2017). Communicating environmental science beyond academia: Stylistic patterns of newsworthiness in popular science journalism. *Discourse & Communication*, 11(1), 69–88.

Monachon, B. (2021). Le choix des cadrages analogiques est-il adapté à la vulgarisation scientifique par la presse ? L'exemple de la lutte biologique. Master's Thesis, Université de Toulon.

Monachon, B. (forthcoming). Le traitement des problématiques environnementales par la Presse Quotidienne Régionale (PQR) : analyse de données textuelles en agroécologie. PhD Thesis, Université de Toulon.

Mondada, L. (1995). La construction discursive des objets de savoir dans l'écriture de la science. *Réseaux*, 71, 55–77.

Morin, E. (1990). *Science avec conscience*. Le Seuil, Paris.

Morin, E. (2005). *Introduction à la pensée complexe*. Le Seuil, Paris.

Mortureux, M.-F. (1988). La vulgarisation scientifique : parole médiane ou dédoublée. In *Vulgariser la science. Le procès de l'ignorance*, Jacobi, D. and Schiele, B. (eds). Champ Vallon, Seyssel, 118–148.

Mortureux, M.-F. (1993). Paradigmes désignationnels. *Semen*, 8 [Online]. Available at: http://journals.openedition.org/semen/4132 [Accessed 22 February 2022].

Mortureux, M.-F. (1994). Traitement des termes et "simplification" dans le discours de vulgarisation scientifique. In *Cahiers du français contemporain. Simple – simplification*, volume 1, Francoise, L. (ed.). ENS, Lyon, 213–225.

Murphy, M.-L. (2006). Antonyms as lexical constructions: Or, why paradigmatic construction is not an oxymoron. *Constructions*, 1(8), 1–37.

Murphy, M.-L., Paradis, C., Willners, C., Jones, S. (2009). Discourse functions of antonymy: A crosslinguistic investigation of Swedish and English. *Journal of Pragmatics*, 41(11), 2159–2184.

Myers, G. (2003). Discourse studies of scientific popularisation: Questioning the boundaries. *Discourse Studies*, 5(2), 265–279.

Naït-Bouda, F. and Pélissier, N. (2021). Le journalisme scientifique : un sous-espace professionnel en dé-différenciation. *Congrès SFSIC Grenoble 2021 – Société et espaces en mouvement*, Grenoble, June.

Nelkin, D. (1986). *Selling Science. How the Press Covers Science and Technology*. W.H. Freeman, New York.

Nisbet, M.C. (2009). The ethics of framing science. In *Communicating Biological Sciences: Ethical and Metaphorical Dimensions*, Nerlich, B., Elliott, R., Larson, B. (eds). Ashgate, Burlington, 51–74.

Nisbet, M.C. and Fahy, D. (2015). The need for knowledge-based journalism in politicized science debates. *The Annals of the American Academy of Political and Social Science*, 658(1), 223–234.

Nisbet, M.C. and Lewenstein, B. (2002). Biotechnology and the American media: The policy process and the elite press, 1970 to 1999. *Science Communication*, 23(4), 359–391.

Oliveira, I. (2009). *Nature et fonctions de la métaphore en science : l'exemple de la cardiologie*. L'Harmattan.

Oreskes, N. and Conway, E. (2012). *Les Marchands de doute*. Le Pommier, Paris.

Pailliart, I., Schiele, B., Jacobi, D., Le Marec, J. (2005). *La publicisation de la science. Exposer, communiquer, débattre, publier, vulgariser*. PUG, Grenoble.

Pal, K.K. and McSpadden Gardener, B. (2006). Biological control of plant pathogens. *The Plant Health Instructor*, 2, 1117–1142 [Online]. Available at: https://www.apsnet.org/edcenter/disimpactmngmnt/topc/Documents/PHI-BiologicalControl.pdf [Accessed 24 February 2022].

Parrenin, F. and Vargas, E. (2020). Biodiversité et changement climatique : entre discours du spécialiste et discours vulgarisé. *Les Carnets du Cediscor*, 15 [Online]. Available at: http://journals.openedition.org/cediscor/2817 [Accessed 24 February 2022].

Pearson, J. (1998). *Terms in Context*. John Benjamins, Amsterdam/Philadelphia.

Pecman, M. (2012). Tentativeness in term formation: A study of neology as a rhetorical device in scientific papers. *Neology in Specialized Communication*, 18(1), 27–58.

Pecman, M. (2014). Variation as a cognitive device: How scientists construct knowledge through term formation. *Terminology*, 20(1), 1–24.

Pecman, M. (2018). *Langue et construction de connaisSENSes. Energie lexico-discursive et potentiel sémiotique des sciences*. L'Harmattan, Paris.

Pecman, M. and Kübler, N. (2022). Text genres and terminology. In *Theoretical Approaches to Terminology From the Other Side of the Looking Glass*, Faber, P. and L'Homme, M.-C. (eds). John Benjamins, Amsterdam/Philadelphia.

Pélissier, N. and Maas, E. (eds) (2014). *Vers une culture médi@tic ? Médias, journalisme et espace public à l'épreuve de la numérisation*. L'Harmattan, Paris.

Peters, H.P. (2012). Le journalisme scientifique : "médier" la relation entre science et société. *Les Cahiers du Journalisme*, 24.

Petit, M. (2010). Le discours spécialisé et le spécialisé du discours : repères pour l'analyse du discours en anglais de spécialité. *E-rea*, 8(1) [Online]. Available at: http://journals.openedition.org/erea/1400 [Accessed 22 February 2022].

Petit, G. (2012). Pour un réexamen de la notion de dénomination. *Langue française*, 174(2), 27–44.

Peynaud, C. (2014). Définition et explication dans la presse généraliste aux États-Unis : de la médiation du discours expert à la création d'un intertexte médiatique. *ASp, Médiation du discours expert : approches linguistiques*, 21–43.

Peynaud, C. (2016). Fukushima and its treatment in the press. *Progress in Industrial Ecology, An International Journal*, 10(2/3), 209–228.

Picton, A. (2009). Diachronie en langue de spécialité. Définition d'une méthode linguistique outillée pour repérer l'évolution des connaissances en corpus. Un exemple appliqué au domaine spatial. PhD Thesis, Université Toulouse 2.

Picton, A. (2018). Terminologie outillée et diachronie : éléments de réflexion autour d'une réconciliation. *ASp*, 74 [Online]. Available at: http://journals.openedition.org/asp/5255 [Accessed 22 February 2022].

Picton, A., Condamines, A., Humbert-Droz, J. (2021). Analyse diachronique du processus de determinologisation. Une réflexion en diachronie courte en physique des particules. *Cahiers de Lexicologie*, 193–225.

Priest, S. (2010). Coming of age in the academy? The status of our emerging field. *Jcom*, 9(3), C06.

Pruvost, J. and Sablayrolles, J.F. (2003). *Les néologismes*. PUF, Paris.

Raichvarg, D. (2016). Vulgarisation versus culture scientifique. Retour vers le futur. In *Histoire de la culture scientifique en France. Institutions et acteurs*, Poirrier, P. (ed.). Éditions universitaires de Dijon, 11–19.

Raichvarg, D. and Jacques, J. (1991). *Savants et ignorants : une histoire de la vulgarisation des sciences*. Le Seuil, Paris.

Rakotonoelina, F. (2014a). Perméabilité des frontières entre l'ordinaire et le spécialisé dans les genres et les discours. *Les Carnets du Cediscor*, 12.

Rakotonoelina, F. (2014b). L'e-conférence écologiste sur les sites web d'(in-) formation : entre genre ordinaire et genre spécialisé. *Les Carnets du Cediscor*, 12, 75–96.

Rastier, F. (2004). Enjeux épistémologiques de la linguistique de corpus. *Texto ! Rubrique Dits et inédits* [Online]. Available at: http://www.revue-texto.net/Inedits/Rastier/Rastier_Enjeux.html [Accessed 22 February 2022].

Rastier, F. (2011). *La mesure et le grain. Sémantique de corpus*. Honoré Champion, Paris.

Rebeyrolle, J. (2014). Sémantique des noms dans les structures énumératives. In *Les Études françaises aujourd'hui*, Vinaver-Kovic, M. and Stanojevic, V. (eds). Faculty of Philology, University of Belgrade, Belgrade, 97–107.

Rebeyrolle, J. and Tanguy, L. (2000). Repérage automatique de structures linguistiques en corpus : le cas des énoncés définitoires. *Cahiers de Grammaire*, 25, 153–174.

Reboul-Touré, S. (2004). Écrire la vulgarisation scientifique aujourd'hui. *Colloque Sciences, Médias et Société*. ENS-LSH, Lyon [Online]. Available at: http://sciences-medias.ens-lyon.fr/article.php3?id_article=65 [Accessed 22 February 2022].

Reboul-Touré, S. (2020). La biodiversité : un mot-témoin pour l'analyse du discours. *Les Carnets du Cediscor*. doi: 10.4000/cediscor.2662. hal-02898357.

Resche, C. (2001). Réflexions sur la frontière entre langue générale et langue spécialisée. *L'Anglais de Spécialité en France*, Geras, 37–46.

Riegel, M. (1987). Définition directe et indirecte dans le langage ordinaire : les énoncés définitoires copulatifs. *Langue française*, 73, 29–53.

Ringoot, R. and Rochard, Y. (2005). Proximité éditoriale : normes et usages des genres journalistiques. *Mots. Les langages du politique*, 77 [Online]. Available at: http://journals.openedition.org/mots/162 [Accessed 22 February 2022].

Ringoot, R. and Utard, J.-M. (eds) (2005). *Le journalisme en invention. Nouvelles pratiques, nouveaux acteurs*. Presses Universitaires de Rennes, Rennes.

Roche, C., Calberg-Challot, M., Damas, L., Rouard, P. (2009). Ontoterminology: A new paradigm for terminology. *International Conference on Knowledge Engineering and Ontology Development*, Madeira, 321–326.

Rogers, M. (1997). Synonymy and equivalence in special-language texts. A case study in german and english texts on genetic engineering. In *Text Typology and Translation*, Trosborg, A. (ed.). John Benjamins, Amsterdam/Philadelphia, 217–245.

Roqueplo, P. (1974). *Le partage du savoir*. Le Seuil, Paris.

Rossetti, A., Cadwell, P., O'Brien, S. (2020). "The terms and conditions came back to bite": Plain language and online financial content for older adults. In *HCI International 2020 – Late Breaking Papers: Universal Access and Inclusive Design*, 699–711.

Rouquette, S. (ed.) (2011). *Sciences et médias*. CNRS, Paris [Online]. Available at: http://books.openedition.org/editionscnrs/15262 [Accessed 22 February 2022].

Roussary, A., Busca, B., Salles, D., Dumont, A., Carpy-Goulard, F. (2013). Pratiques phytosanitaires en agriculture et environnement. Des tensions irréductibles ? *Économie rurale*, 333, 67–80.

Ruellan, D. (1992). Le professionnalisme du flou. *Réseaux*, 10(51), 25–37.

Schiele, B. (1983). Les enjeux cachés de la vulgarisation scientifique. *Communication Information*, 5(2/3), 156–185.

Schiele, B. (2005). Publiciser la science ! Pour quoi faire ? Revisiter la notion de culture scientifique et technique. In *La publicisation de la science*, Paillart, I. (ed.). PUG, Grenoble.

Schoenfeld, A.C. (1980). Newspersons and the environment today. *Journalism Quarterly*, 57(3), 456–462.

Secko, D. (2012a). Un sujet, deux articles : leçons d'une expérience journalistique sur la génomique. *Les Cahiers du Journalisme*, 24.

Secko, D., Amend, E., Friday, T. (2013). Four models of science journalism: A synthesis and practical assessment. *Journalism Practice*, 7(1), 62–80.

Sicard, M.-N. (1997). Pratiques journalistiques et enjeux de la communication scientifique et technique. *Revue Hermès*, 21, 149–155.

Sicard, M.-N. (2011). Les journalistes scientifiques : régulateurs dans un espace d'opinion. In *Sciences et médias*, Rouquette, S. (ed.). CNRS, Paris [Online]. Available at: http://books.openedition.org/editionscnrs/15286 [Accessed 22 February 2022].

Sinclair, J. (2000). Lexical grammar. *Naujoji Metodologija*, 24, 191–203.

Sitri, F. (2003). *L'objet du débat*. Presses Sorbonne Nouvelle, Paris.

Stocklmayer, S. and Gilbert, J.-K. (2002). New experiences and old knowledge: Towards a model for the personal awareness of science and technology. *International Journal of Science Education*, 24(8), 835–858.

Stosic, D. and Rebeyrolle, J. (2021). Les noms sous-spécifiés en français : du lexique au discours. Présentation. *Corela*, HS-34 [Online] Available at: http://journals.openedition.org/corela/13467 [Accessed 13 January 2022].

Temmerman, R. (2000). *Towards New Ways of Terminology Description. The Sociocognitive Approach*. John Benjamins, Amsterdam/Philadelphia.

Thouin, M. and Lapointe, P. (2016). La médiation scientifique ouverte. Tribune. *ACFAS Magazine* [Online]. Available at: https://www.acfas.ca/publications/magazine/2016/12/mediation-scientifique-ouverte [Accessed 22 February 2022].

Tolédano, V. and Candel, D. (2006). Mouvance terminologique, l'espace d'une courte diachronie, dans les dictionnaires et dans la presse : le cas de "mondialisation/globalisation". In *Aspects diachroniques du vocabulaire*, Candel, D. and Gaudin, F. (eds). Presses Universitaires de Rouen et du Havre, Rouen, 157–183.

Tournier, M. (1994). Petite suite en vert. *Mots*, June(39), 114–116.

Treise, D. and Weingold, L.F. (2002). Advancing science communication. *Science Communication*, 23(3), 310–322.

Trench, B. and Bucchi, M. (2010). Science communication, an emerging discipline. *JCOM*, 9(03), C03. doi: 10.22323/2.09030303.

Valenti, J.M. and Tavana, G. (2005). Report: Continuing science education for environmental journalists and science writers. *Science Communication*, 27(2), 300–310.

Van der Yeught, M. (2019). Une épistémologie et une théorie des langues de spécialité au service de solutions pour le secteur LANSAD. In *Épistémologie à usage didactique. Langues de spécialité*, Chaplier, C. and O'Connell, A.-M. (eds). L'Harmattan, 53–95.

Van Dijk, T. (1999). Context models in discourse processing. In *The Construction of Mental Representations During Reading*, Van Oostendorp, H. and Goldman, S. (eds). Lawrence Erlbaum, London, 123–148.

Van Dijk, T. (2003). The discourse-knowledge interface. In *Critical Discourse Analysis: Theory and Interdisciplinarity*, Wodak, R. and Weiss, G. (eds). Palgrave Macmillan, London, 85–109.

Vanderpooten, G. (2020). *Imaginer le monde de demain Le rôle positif des médias*. Actes Sud, Arles.

Varela, F. (1988). *Invitation aux sciences cognitives*. Le Seuil, Paris.

Véron, É. (1997). Entre l'épistémologie et la communication. *Revue Hermès*, 21(1), 23–32.

Vintar, S. and Grčić Simeunović, L. (2017). Definition frames as language-dependent models of knowledge transfer. *Fachsprache*, 1(2), 43–58.

Weingart, P. (2001). Die Stunde der Wahrheit? Zum Verhältnis der Wissenschaft zu Politik, Wirtschaft und Medien in der Wissensgesellschaft. Velbrück, Weilerswist.

Whitley, R. (1985). Editorial preface. In *Expository Science: Forms and Functions of Popularisation*, Shinn, T. and Whitley, R. (eds). Reidel, Dortrecht.

Williams Camus, J. (2009). Metaphors of cancer in scientific popularization articles in the British press. *Discourse Studies*, 11(4), 465–495.

Winsten, J.A. (1985). Science and the media: The boundaries of truth. *Health Affairs (Project Hope)*, 4(1), 5–23.

Wolton, D. (1997). De la vulgarisation à la communication. *Revue Hermès*, 21, 9–14.

Wolton, D. (2011). De la vulgarisation à la communication. In *Sciences et médias*, Rouquette, S. (ed.). CNRS Éditions. doi: 10.4000/books.editionscnrs.15277.

Zappalà, A. (1997). La médecine médiatisée : entre la médicalisation du social et la socialisation de la science. *Revue Hermès*, 21, 181–190.

Index

O, P

R, S

T, U, V

Other titles from

iSTE

in

Science, Society and New Technologies

2024

BARTHES Angela, CIBIEN Catherine, ROMAGNY Bruno
Biosphere Reserves and Sustainable Development Goals 1: Scientific and Practical Educational Issues in the Mediterranean
(Territory Development Set – Volume 3)
Biosphere Reserves and Sustainable Development Goals 2: Issues, Tensions, Processes and Governance in the Mediterranean
(Territory Development Set – Volume 4)

2023

APCHAIN Thomas
Favela Tours: Building Otherness in the Favelas of Rio de Janeiro
(Tourism and Mobility Systems Set – Volume 6)

CARDON Alain, EL HAMI Abdelkhalak
Fundamental Generation Systems: Computer Science and Artificial Consciousness, the Informational Field of Generation of the Universe, the Sixth Sense of Living Beings
(Digital Sciences Set – Volume 4)

CURULLI Irene G., IKIZ KAYA Deniz, KHAEFI Arghavan
Heritage Education for Climate Action
(Research in Architectural Education Set – Volume 2)

ELAMÉ Esoh
The Sustainable City in Africa Facing the Challenge of Liquid Sanitation
(Territory Development Set – Volume 2)

JURCZENKO Emmanuel
Climate Investing: New Strategies and Implementation Challenges

LYUBAREVA Inna, WALDECK Roger
Diversity of Methodological Approaches in Social Sciences: Example of the Analysis of Media and Online Information
(Modeling Methodologies in Social Sciences Set – Volume 3)

PARET Dominique, CRÉGO Pierre, SOLÈRE Pauline
Smart Patches: Biosensors, Graphene and Intra-Body Communications

VENTRE Daniel, GUILLOT Philippe
Electronic Communication Interception Technologies and Issues of Power

VENTRE Daniel, LOISEAU Hugo
Cybercrime during the SARS-CoV-2 Pandemic (2019-2022): Evolutions, Adaptations, Consequences
(Cybersecurity Set – Volume 3)

2022

AIT HADDOU Hassan, TOUBANOS Dimitri, VILLIEN Philippe
Ecological Transition in Education and Research

CARDON Alain
Information Organization of The Universe and Living Things: Generation of Space, Quantum and Molecular Elements, Coactive Generation of Living Organisms and Multiagent Model
(Digital Sciences Set – Volume 3)

CAULI Marie, FAVIER Laurence, JEANNAS Jean-Yves
Digital Dictionary

DAVERNE-BAILLY Carole, WITTORSKI Richard
Research Methodology in Education and Training: Postures, Practices and Forms
(Education Set – Volume 12)

ELAMÉ Esoh
Sustainable Intercultural Urbanism at the Service of the African City of Tomorrow
(Territory Development Set – Volume 1)

FLEURET Sébastien
A Back and Forth Between Tourism and Health: From Medical Tourism to Global Health
(Tourism and Mobility Systems Set – Volume 5)

KAMPELIS Nikos, KOLOKOTSA Denia
Smart Zero-energy Buildings and Communities for Smart Grids
(Engineering, Energy and Architecture Set – Volume 9)

2021

BARDIOT Clarisse
Performing Arts and Digital Humanities: From Traces to Data
(Traces Set – Volume 5)

BENSRHAIR Abdelaziz, BAPIN Thierry
From AI to Autonomous and Connected Vehicles: Advanced Driver-Assistance Systems (ADAS)
(Digital Sciences Set – Volume 2)

BÉRANGER Jérôme
Societal Responsibility of Artificial Intelligence: Towards an Ethical and Eco-responsible AI
(Technological Prospects and Social Applications Set – Volume 4)

CORDELIER Benoit, GALIBERT Oliver
Digital Health Communications
(Technological Prospective and Social Applications Set – Volume 5)

DOUAY Nicolas, MINJA Michael
Urban Planning for Transitions

GALINON-MÉLÉNEC Béatrice
The Trace Odyssey 1: A Journey Beyond Appearances
(Traces Set – Volume 4)

HENRY Antoine
Platform and Collective Intelligence: Digital Ecosystem of Organizations

KOLOKOTSA Denia, KAMPELIS Nikos
Smart Buildings, Smart Communities and Demand Response
(Engineering, Energy and Architecture Set – Volume 8)

LE LAY Stéphane, SAVIGNAC Emmanuelle, LÉNEL Pierre, FRANCES Jean
The Gamification of Society
(Research, Innovative Theories and Methods in SSH Set – Volume 2)

RADI Bouchaïb, EL HAMI Abdelkhalak
Optimizations and Programming: Linear, Non-linear, Dynamic, Stochastic and Applications with Matlab
(Digital Sciences Set – Volume 1)

2020

ALAKTIF Jamila, CALLENS Stéphane
Migration and Climate Change: From the Emergence of Human Cultures to Contemporary Management in Organizations

BARNOUIN Jacques
The World's Construction Mechanism: Trajectories, Imbalances and the Future of Societies
(Interdisciplinarity Between Biological Sciences and Social Sciences Set – Volume 4)

ÇAĞLAR Nur, CURULLI Irene G., SIPAHIOĞLU Işıl Ruhi, MAVROMATIDIS Lazaros E.
Thresholds in Architectural Education
(Engineering, Energy and Architecture Set – Volume 7)

DUBOIS Michel J.F.
Humans in the Making: In the Beginning was Technique
(Social Interdisciplinarity Set – Volume 4)

ETCHEVERRIA Olivier
The Restaurant, A Geographical Approach: From Invention to Gourmet Tourist Destinations
(Tourism and Mobility Systems Set – Volume 3)

GREFE Gwenaëlle, PEYRAT-GUILLARD Dominique
Shapes of Tourism Employment: HRM in the Worlds of Hotels and Air Transport
(Tourism and Mobility Systems Set – Volume 4)

JEANNERET Yves
The Trace Factory
(Traces Set – Volume 3)

KATSAFADOS Petros, MAVROMATIDIS Elias, SPYROU Christos
Numerical Weather Prediction and Data Assimilation
(Engineering, Energy and Architecture Set – Volume 6)

MARTI Caroline
Cultural Mediations of Brands: Unadvertization and Quest for Authority
(Communication Approaches to Commercial Mediation Set – Volume 1)

MAVROMATIDIS Lazaros
Climatic Heterotopias as Spaces of Inclusion: Sew Up the Urban Fabric
(Research in Architectural Education Set – Volume 1)

MOURATIDOU Eleni
Re-presentation Policies of the Fashion Industry: Discourse, Apparatus and Power
(Communication Approaches to Commercial Mediation Set – Volume 2)

SCHMITT Daniel, THÉBAULT Marine, BURCZYKOWSKI Ludovic
Image Beyond the Screen: Projection Mapping

VIOLIER Philippe, with the collaboration of TAUNAY Benjamin
The Tourist Places of the World
(Tourism and Mobility Systems Set – Volume 2)

2019

BRIANÇON Muriel
The Meaning of Otherness in Education: Stakes, Forms, Process, Thoughts and Transfers
(Education Set – Volume 3)

DESCHAMPS Jacqueline
Mediation: A Concept for Information and Communication Sciences
(Concepts to Conceive 21st Century Society Set – Volume 1)

DOUSSET Laurent, PARK Sejin, GUILLE-ESCURET Georges
Kinship, Ecology and History: Renewal of Conjectures
(Interdisciplinarity Between Biological Sciences and Social Sciences Set – Volume 3)

DUPONT Olivier
Power: A Concept for Information and Communication Sciences
(Concepts to Conceive 21st Century Society Set – Volume 2)

FERRARATO Coline
Prospective Philosophy of Software: A Simondonian Study
(Technological Prospects and Social Applications Set – Volume 3)

GUAAYBESS Tourya
The Media in Arab Countries: From Development Theories to Cooperation Policies

HAGÈGE Hélène
Education for Responsibility
(Education Set – Volume 4)

LARDELLIER Pascal
The Ritual Institution of Society
(Traces Set – Volume 2)

LARROCHE Valérie
The Dispositif: A Concept for Information and Communication Sciences
(Concepts to Conceive 21st Century Society Set – Volume 3)

LATERRASSE Jean
Transport and Town Planning: The City in Search of Sustainable Development

LELEU-MERVIEL Sylvie, SCHMITT Daniel, USEILLE Philippe
From UXD to LivXD: Living eXperience Design

LENOIR Virgil Cristian
Ethically Structured Processes
(Innovation and Responsibility Set – Volume 4)

LOPEZ Fanny, PELLEGRINO Margot, COUTARD Olivier
Local Energy Autonomy: Spaces, Scales, Politics
(Urban Engineering Set – Volume 1)

METZGER Jean-Paul
Discourse: A Concept for Information and Communication Sciences
(Concepts to Conceive 21st Century Society Set – Volume 4)

MICHA Irini, VAIOU Dina
Alternative Takes to the City
(Engineering, Energy and Architecture Set – Volume 5)

PÉLISSIER Chrysta
Learner Support in Online Learning Environments

PIETTE Albert
Theoretical Anthropology or How to Observe a Human Being
(Research, Innovative Theories and Methods in SSH Set – Volume 1)

PIRIOU Jérôme
The Tourist Region: A Co-Construction of Tourism Stakeholders
(Tourism and Mobility Systems Set – Volume 1)

PUMAIN Denise
Geographical Modeling: Cities and Territories
(Modeling Methodologies in Social Sciences Set – Volume 2)

WALDECK Roger
Methods and Interdisciplinarity
(Modeling Methodologies in Social Sciences Set – Volume 1)

2018

BARTHES Angela, CHAMPOLLION Pierre, ALPE Yves
Evolutions of the Complex Relationship Between Education and Territories
(Education Set – Volume 1)

BÉRANGER Jérôme
The Algorithmic Code of Ethics: Ethics at the Bedside of the Digital Revolution
(Technological Prospects and Social Applications Set – Volume 2)

DUGUÉ Bernard
Time, Emergences and Communications
(Engineering, Energy and Architecture Set – Volume 4)

GEORGANTOPOULOU Christina G., GEORGANTOPOULOS George A.
Fluid Mechanics in Channel, Pipe and Aerodynamic Design Geometries 1
(Engineering, Energy and Architecture Set – Volume 2)
Fluid Mechanics in Channel, Pipe and Aerodynamic Design Geometries 2
(Engineering, Energy and Architecture Set – Volume 3)

GUILLE-ESCURET Georges
Social Structures and Natural Systems: Is a Scientific Assemblage Workable?
(Social Interdisciplinarity Set – Volume 2)

LARINI Michel, BARTHES Angela
Quantitative and Statistical Data in Education: From Data Collection to Data Processing
(Education Set – Volume 2)

LELEU-MERVIEL Sylvie
Informational Tracking
(Traces Set – Volume 1)

SALGUES Bruno
Society 5.0: Industry of the Future, Technologies, Methods and Tools
(Technological Prospects and Social Applications Set – Volume 1)

TRESTINI Marc
Modeling of Next Generation Digital Learning Environments: Complex Systems Theory

2017

ANICHINI Giulia, CARRARO Flavia, GESLIN Philippe, GUILLE-ESCURET Georges
Technicity vs Scientificity: Complementarities and Rivalries (Interdisciplinarity Between Biological Sciences and Social Sciences Set – Volume 2)

DUGUÉ Bernard
Information and the World Stage: From Philosophy to Science, the World of Forms and Communications (Engineering, Energy and Architecture Set – Volume 1)

GESLIN Philippe
Inside Anthropotechnology: User and Culture Centered Experience (Social Interdisciplinarity Set – Volume 1)

GORIA Stéphane
Methods and Tools for Creative Competitive Intelligence

KEMBELLEC Gérald, BROUDOUX Evelyne
Reading and Writing Knowledge in Scientific Communities: Digital Humanities and Knowledge Construction

MAESSCHALCK Marc
Reflexive Governance for Research and Innovative Knowledge (Responsible Research and Innovation Set - Volume 6)

PARK Sejin, GUILLE-ESCURET Georges
Sociobiology vs Socioecology: Consequences of an Unraveling Debate (Interdisciplinarity Between Biological Sciences and Social Sciences Set – Volume 1)

PELLÉ Sophie
Business, Innovation and Responsibility (Responsible Research and Innovation Set – Volume 7)

2016

ANDRÉ Michel, SAMARAS Zissis
Energy and Environment
(Research for Innovative Transports Set – Volume 1)

BLANQUART Corinne, CLAUSEN Uwe, JACOB Bernard
Towards Innovative Freight and Logistics
(Research for Innovative Transports Set – Volume 2)

COHEN Simon, YANNIS George
Traffic Management
(Research for Innovative Transports Set – Volume 3)

EL FALLAH SEGHROUCHNI Amal, ISHIKAWA Fuyuki, HÉRAULT Laurent, TOKUDA Hideyuki
Enablers for Smart Cities

GIANNI Robert
Responsibility and Freedom: The Ethical Realm of RRI
(Responsible Research and Innovation Set – Volume 2)

GRUNWALD Armin
The Hermeneutic Side of Responsible Research and Innovation
(Responsible Research and Innovation Set – Volume 5)

LENOIR Virgil Cristian
Ethical Efficiency: Responsibility and Contingency
(Responsible Research and Innovation Set – Volume 1)

PELLÉ Sophie, REBER Bernard
From Ethical Review to Responsible Research and Innovation
(Responsible Research and Innovation Set – Volume 3)

REBER Bernard
Precautionary Principle, Pluralism and Deliberation: Science and Ethics
(Responsible Research and Innovation Set – Volume 4)

TORRENTI Jean-Michel, LA TORRE Francesca
Materials and Infrastructures 1
(Research for Innovative Transports Set – Volume 5A
Materials and Infrastructures 2
(Research for Innovative Transports Set – Volume 5B)

VENTRE Daniel
Information Warfare – 2nd edition

YANNIS George, COHEN Simon
Traffic Safety
(Research for Innovative Transports Set – Volume 4)

2015

BRONNER Gérald
Belief and Misbelief Asymmetry on the Internet

2009

VENTRE Daniel
Information Warfare

Printed and bound by CPI Group (UK) Ltd, Croydon, CR0 4YY

13/09/2026

14961039-0002